Finite Element Methods for
Nonlinear Optical Waveguides

T0239810

Advances in Nonlinear Optics

A series edited by Anthony F. Garito, *University of Pennsylvania, USA* and François Kajzar, *DEIN, CEN de Saclay, France*

This book is part of a series. The publisher will accept continuation orders which may be cancelled at any time and which provide for automatic billing and shipping of each title in the series upon publication. Please write for details.

Finite Element Methods for Nonlinear Optical Waveguides

Xin-Hua Wang
Department of Electrical and Electronic Engineering
University of Melbourne
Australia

CRC Press
Taylor & Francis Group
Boca Raton London New York

CRC Press is an imprint of the
Taylor & Francis Group, an **informa** business

First published 1995 by Gordon and Breach Science Publishers

Published 2019 by CRC Press
Taylor & Francis Group
6000 Broken Sound Parkway NW, Suite 300
Boca Raton, FL 33487-2742

© 1995 by Taylor & Francis Group, LLC
CRC Press is an imprint of Taylor & Francis Group, an Informa business

First issued in paperback 2019

No claim to original U.S. Government works

ISBN-13: 978-0-367-45595-8 (pbk)
ISBN-13: 978-2-88449-048-1 (hbk)

Visit the Taylor & Francis Web site at
http://www.taylorandfrancis.com

and the CRC Press Web site at
http://www.crcpress.com

British Library Cataloguing in Publication Data

Wang, Xin-Hua
   Finite Element Methods for Nonlinear
   Optical Waveguides. – (Advances in
   Nonlinear Optics Series, ISSN 1068-672X;
   Vol.2)
   I. Title   II. Series
   621.36

CONTENTS

INTRODUCTION TO THE SERIES

Advances in Nonlinear Optics is a series of original monographs and collections of review papers written by leading specialists in quantum electronics. It covers recent developments in different aspects of the subject including fundamentals of nonlinear optics, nonlinear optical materials both organic and inorganic, nonlinear optical phenomena such as phase conjugation, harmonic generation, optical bistability, fast and ultrafast processes, waveguided nonlinear optics, nonlinear magneto-optics and waveguiding integrated devices.

The series will complement the international journal *Nonlinear Optics: Principles, Materials, Phenomena and Devices* and is foreseen as material for teaching graduate and undergraduate students, for people working in the field of nonlinear optics, for device engineers, for people interested in a special area of nonlinear optics and for newcomers.

INTRODUCTION TO THE SERIES

PREFACE

Much world-wide interest has been paid to nonlinear optical waveguides in the last fifteen years or so as the intensity-dependent dielectric properties of nonlinear optical waveguides have many potential applications in all-optical signal processing devices. The accurate analysis of such waveguide structures strictly needs the solution of a vectorial and nonlinear partial-differential equation. Thus it is necessary to employ a powerful numerical approach such as the finite element method. The intense activity in the study of nonlinear optical waveguide has resulted in the general need for a systematic procedure and powerful software tools. However, finite-element programming requires expertise in mathematics, physics and/or engineering, plus great care and patience. Moreover, even if a finite-element program is available for use, the preparation of a suitable mesh for a complicated structure is usually a formidable task if done manually.

This book is intended to provide a robust procedure for the systematic investigation of nonlinear optical waveguides. Specifically, correct simulations, reliable and efficient computations and the study of fundamental phenomena are of prime concern. For the reader to be able to make the best use of this book, a powerful, self-contained and user-friendly software package ANOWS (Advanced Nonlinear Optical Waveguide Simulator) written in Fortran 77 source code is attached to the book for general use.

The book is primarily aimed at researchers already engaged in or wishing to enter the field of nonlinear-optical waveguides. However, the methodologies contained in the book and the software package attached can equally be applied to linear optical waveguides as well as microwave waveguides which do not support TEM modes. Hence they should be useful for scientists and engineers interested in linear and nonlinear waveguide devices. The software package and portions of the text can also be used for graduate courses dealing with waveguide modes.

The book starts with the weak formulation of the full vectorial electromagnetic wave equation. Then the operator equation is approximated by the method of moments and solved by the finite element method. To relax the restrictions on admissible functions imposed by discontinuity and boundary conditions, an extended operator method is fully described. The nonlinear algebraic matrix equation obtained is solved in a linear way by a nonlinear iteration scheme armed with a rather effective acceleration technique. The convergence behaviour of the iterations is monitored by established *a posteriori* error estimates. To reduce errors and the time involved in data preparation, an automatic mesh generation scheme is presented, including a novel technique for node-renumbering higher-order meshes.

Among various nonlinear optical waveguides, planar structures have been extensively investigated with quasi-2D (using one transverse spatial variable) formulations. In this book it is shown that, whereas weak nonlinear effects are well simulated in quasi-2D, strong nonlinear actions in planar structures must be simulated in quasi-3D (using two transverse spatial variables). In quasi-2D formulations, a saturable model of the nonlinear permittivity is required only for physical reasons. In quasi-3D formulations, it is shown that for both the physical and the mathematical requirements saturation effects must be incorporated into the nonlinear permittivity model when simulating strong self-focussing. The results of the two formulations are compared qualitatively. How to characterise planar nonlinear optical waveguides, whether in terms of total guided power or power per unit length, is discussed in detail.

Novel solution algorithms are presented for computing nonlinear optical waveguides exhibiting bistability phenomena. Both stable and unstable solutions can be computed by a neat reformulation in terms of either the electric or the magnetic field. A normalisation procedure is also incorporated.

The analysis of nonlinear optical waveguides with a full vectorial formulation is accurate but expensive. A scalar approximation procedure for weakly-guiding structures is established. The quantitative comparison of its results and those of the vectorial formulation is given for both planar and channel structures.

The stability analysis of nonlinear modes is crucial to the application of nonlinear modal methods. An affordable robust procedure dedicated to scalar nonlinear wave propagation in three dimensions is developed based on the FEM-FDM (finite-element method plus finite-difference method). With this procedure nonlinear modal methods are justified on the basis of the propagation properties of nonlinear modes (spatial solitons) and quasi-modes; also the bistability phenomenon predicted by the modal method is confirmed numerically. In the context of all-optical switching, it is shown how to launch a field distribution which remains well confined during propagation.

Nonlinear coupled waveguides are most useful devices for all-optical signal processing. Three frequently-used techniques for analysing linear and nonlinear optical couplers are described, namely, the supermode superposition technique, coupled-mode theory and the propagation method. Linear and nonlinear coupled-mode equations are derived by using a reciprocity approach. These equations are very general and can be applied to vector mode coupling in anisotropic and lossy media provided that the media are non-magnetic and z-independent. It is demonstrated how propagation methods combined with modal methods can be applied to linear and nonlinear coupled waveguides via a numerical example. It is also shown how to improve the computational efficiency.

In this book, several examples of nonlinear optical waveguides are investigated. The results show that these structures may well find applications in all-optical switching, low-threshold devices and bistability or even multistability devices.

Finally, the contents, documentation, usage and examples of the software package ANOWS are given in the appendices. The author hopes the source code will be widely used by researchers but can take no responsibility for the results generated or for the consequences of any modifications made to the code.

Acknowledgements

This book is based on the author's Ph.D. research work carried out in the Department of Electrical and Computer Systems Engineering, Monash University, Australia, together with his subsequent journal publications. It was written while the author was a research fellow in the Photonics Research Laboratory (PRL), Department of Electrical and Electronic Engineering, The University of Melbourne, Australia.

I am particularly indebted to my former Ph.D. supervisor, Dr. G. K. Cambrell. He has given his time generously during my Ph.D. study and for the completion of this book. Many substantial discussions with him have always been a source of inspiration resulting in a stimulating research environment. He checked the whole book thoroughly and offered many invaluable suggestions for improving its clarity and presentation. His influence has pervaded the entire book, particularly *extended operator theories, function spaces* and *dyadic analysis*. Without his sincere help and constant encouragement, the publication of this book would have been impossible.

I wish to thank Professors F. Kajzar of Centre d'Études Nucléaires de Saclay, France, and A. Garito of The University of Pennsylvania, USA, for their enthusiasm in including this book in the series. The support of Professors W. A. Brown of Monash University and R. S. Tucker of PRL is warmly acknowledged. Thanks are also due to Professors C. Pask of The Australian Defence Force Academy and G. I. Stegeman of The University of Central Florida, USA, for their most favourable comments on my Ph.D. thesis, which was one of the driving forces for my writing this book. To Ray, my son, who was denied my companionship countless times over the past few years, I owe a special debt. Finally, a very special word of thanks to Liping, my wife, who has sacrificed herself for this research work. Her understanding, support, continual encouragement and tolerance of my absences are greatly appreciated.

To Liping and Ray

Learning without thinking causes haze;
Thinking without learning causes bewilderment.

– Confucius

Chapter 1

Introduction

1.1 Introduction

Electronic communications systems are said to make use of electronic devices with electrons being information carriers for signal processing and transmission. So one may expect optical communications systems to make use of all-optical devices built on photons. The advance in communications from electronic systems to optical fibre systems is one of the greatest achievements in modern technology. Optical communication through fibre cables has several overwhelming advantages over conventional metallic or coaxial cable facilities, such as greater capacity, noise immunity, safety and security, resistance to environmental extremes and low cost in the long run. Optical fibre communications also challenge the "newly-born" satellite communications.

Compared with electronic components, optical devices are much faster (transition times measured in picoseconds) and have a much higher capacity for integration. However, current optical communications systems are not "all-optical" but rather "electronic-optical-electronic". For long-haul communications where repeaters are required, the signal transmitted has to be converted from optical to electronic and then back to optical. Such a tortuous process is not only cost-ineffective but also introduces undesired delay. Moreover, the speed of electronic devices will be an ultimate barrier for increasing the capacity of optical fibre communications systems. Thus, there is an impetus to develop all-optical devices to replace those electronic substitutes. Apart from communications, the speed of electronic devices has proven to be the "bottleneck" of electronic computers. It is hardly imaginable that the speed of the

CPU in an electronic computer can increase by an order of magnitude from its present stage of development. However, all-optical computers are expected to be many thousands of times faster than their electronic "parents".

For optical signal processing based on all-optical switching, bistability and logic, the operation is strictly nonlinear and therefore it is natural to make use of nonlinear optical effects in materials. Nonlinear optics is not new, but its application to all-optical signal processing devices became possible only when very powerful lasers came into being since the nonlinear coefficients of optical materials are usually relatively small. Considerable world interest in making use of such effects started only in the early 1980s to catch up with the rapid development of optical fibre communications technology.

The enhancement of nonlinear optical effects can be achieved by three different approaches: developing more powerful lasers, searching for new optical materials with large nonlinear coefficients and low dissipation, and using efficient guiding geometries. While the development of powerful lasers and the search for new materials are continuing, this book covers only the last approach: using guiding geometries. The use of novel guiding geometries also offers many fascinating device-oriented phenomena which cannot be derived from the other two approaches alone.

Accurate analysis of nonlinear waveguide structures strictly needs solving the vectorial nonlinear partial-differential equation. Nonlinear optical waveguides whose governing equations possess solutions in a closed form are very rare, and thus the analytical approach is rather restrictive. On the contrary, even very complicated problems can be readily solved by a numerical procedure. Once it is set up, it can be applied to a wide class of problems. On the other hand, even if an analytical solution does exist, the results are often too complicated to interpret without numerical computation.

It is well-known that finite element methods provide a home for numerical solutions of partial-differential equations and have many distinct advantages over other numerical methods such as finite difference methods[1] and boundary element methods[2]. Although finite element methods have been successfully applied to the analysis of linear wave-

---

[1]In literature, finite difference methods are distinguished from finite element methods. In fact, the former can be regarded as a special case of the latter.

[2]Even boundary element methods can be regarded as a class of finite element method in which an integral equation is solved for an unknown field on a boundary.

guide structures for many years, the generalisation to nonlinear problems is by no means trivial as nonlinear optical waveguides behave quite differently from their linear counterparts and all terminology such as "modes" needs re-examining and interpreting in a nonlinear situation.

While the search for new nonlinear optical materials is accelerating, it is premature to say whether any particular material will be the ultimate one for nonlinear integrated optics; possibly it may be a member of the organic polymer family. At the present stage it is felt that correct simulations, reliable and efficient computations and the investigation of fundamental phenomena of nonlinear guided waves are more important than sophisticated designs of specific devices so that the challenge of forthcoming materials and device requirements can be met. Also, a robust solution procedure for a variety of structures is of great value. That is the motivation of this book, which may be stated as an old Chinese proverb: "Supplies go ahead of troops arriving!"

In essence, the framework of the above *motivation* has defined the following *objectives* in this book:

- To develop robust algorithms for correct simulations and reliable and efficient computations of nonlinear optical waveguides;

- To investigate fundamental phenomena of nonlinear guided waves and their potential applications;

- To provide a reliable and user-friendly software package for researchers to use.

The area of nonlinear guided waves is rather broad and we solely consider monochromatic light propagating in guided structures consisting of linear and nonlinear self-focussing and/or self-defocussing media, and ignore any harmonic generation occurring.

1.2   Overview

The book starts in Chapter 2 with the time-harmonic wave equation for either the electric or the magnetic field, a second-order partial-differential equation defined over the cross-section of the optical waveguide, together with discontinuity conditions for material interfaces and boundary conditions for the "walls" of the structure. This is expressed as a linear-operator equation in an appropriate function space, ignoring any medium nonlinearity for the time being. The domain of the linear operator incorporates all of the discontinuity and boundary conditions. Next, using duality pairing in the function space, it is shown how to derive a weak formulation of the linear-operator equation which extends its domain to functions which are only piecewise differentiable to the first order. Then the linear operator is shown to be self-adjoint provided the permittivity and permeability dyadics are hermitian.

Since the electromagnetic field in the waveguide cross-section is source-free, and modes of propagation are sought at a given frequency, the linear-operator equation becomes an eigenvalue problem in which eigenpairs, consisting of an eigenvalue and an eigenvector, are sought. It is shown that the eigenpairs are real since the linear operator is self-adjoint, leading to cheaper computation.

The weak form of the linear-operator equation is next reduced to a matrix equation by application of either a projection method or a variational method. No variational functional is required in order to apply a projection method, which in this sense is a more direct technique. In the general method of moments, a common projection method, the unknown electromagnetic field is approximated by a linear combination of suitable expansion functions with unknown coefficients. After substituting into the weak form of the operator equation, the equation residual is forced to zero by "testing" the equation with suitable testing or weighting functions equal in number to the number of unknown coefficients, resulting in a square matrix equation. The testing technique is actually the process of setting to zero the projection of the linear-operator equation residual into the space of testing functions; the inner products involved are called "moments". A special case of the general method of moments is Galerkin's method in which the set of testing functions is the same as the set of expansion functions; this is the method adopted in this book.

To solve the linear-operator equation, expansion and testing functions need to satisfy the associated interface and boundary conditions in general. In this book the linear operator itself is extended to incorporate all of the discontinuity and boundary conditions as well as the partial-differential operator into one entity. The domain of this extended operator now includes functions that do not necessarily satisfy all of the discontinuity and boundary conditions, which enhances the flexibility of the finite element method.

Since the exact modal eigenfunctions are solenoidal, it is necessary to impose a solenoidal constraint on the approximate solutions, otherwise many spurious numerical solutions may appear. Because the construction of a solenoidal vector function subspace has proven to be very difficult for quasi-3D (using two transverse spatial variables) problems, a penalty-function method is employed in this book for imposing the solenoidal constraint. Effectively this changes the second-order partial-differential operator from *curl-curl* type to *curl-curl — grad-div* type, which is a positive-definite operator similar to the negative[3] vector Laplacian operator having a zero-dimensional null-space (for simple geometries). Consequently, the eigenvalues of both the operator equation and the corresponding matrix equation are positive.

The vector finite-element approach is examined in Chapter 3. The expansion and testing functions in Galerkin's method are regarded as vector functions, rather than a set of three scalar functions, for approximating vector fields. Then vector field constraints, such as discontinuity and boundary conditions, can be easily imposed. Although none of these conditions need be satisfied *a priori* when using the extended operator, it is preferable that at least some of the discontinuity and boundary conditions be satisfied explicitly. Details of such vector field constraints for anisotropic media are described in Chapter 3.

To reduce errors and the time involved in data preparation, an automatic mesh generation algorithm for triangular elements is developed in Chapter 4 on the basis of known strategies. The quality of mesh networks produced is then improved by a smoothing procedure called node relaxation, and the bandwidth is minimised by automatic node renumbering. In particular, a new scheme of renumbering higher-order mesh networks is presented.

In selecting solution techniques for the algebraic matrix equation, the nature of the problem is scrutinised and several efficient solution methods are briefly discussed in Chapter 5. As the iterative solution method for *nonlinear* guided waves strongly relies on the solution of *linear* waveguide problems, the software developed is then tested against a linear structure having a known analytical solution. To further justify the iterative solution algorithm, an *a posteriori* error estimate is established. In addition, Chapter 5 introduces an effective acceleration technique to speed up the nonlinear iteration process. The iterative solution algorithm given in Chapter 5 is conventional in that the effective modal index is extracted as the eigenvalue for given guided power. Another iterative solution algorithm, initiated in the present work, is presented

---

[3]Note that for a twice-differentiable vector field V, $\nabla \times (\nabla \times V) - \nabla(\nabla \cdot V) = -\nabla^2 V$.

in Chapter 6.

In Chapter 6, a comparison of our results and those available in the literature is given for a classical channel structure filled with linear and nonlinear media, and a significant discrepancy is observed. The cause of the discrepancy is then identified. Consequently, an important conclusion on how to correctly simulate nonlinear guided waves confined in both transverse directions is reached.

To make power dispersion relations more universal, a normalisation procedure is proposed, by which both the normalised electromagnetic field and the normalised guided power are made dimensionless and independent of the nonlinear coefficient. With this normalisation procedure, the consistency and the efficiency of the electric- and the magnetic-field formulations are examined critically in terms of a nonlinear-film-loaded ion-exchanged-channel structure. Also, the effectiveness of the acceleration technique employed is demonstrated.

For the above two nonlinear structures investigated, one shows an abrupt jump in the power dispersion curve, whereas the other does not. Some factors giving rise to this jump are outlined by comparing the two structures. In particular, it is predicted with success that certain nonlinear strip-loaded channel waveguides possess a jump in their power dispersion curves. In this book, new solution algorithms, valid for both the electric and magnetic field formulation and a wide class of nonlinear mechanisms, are presented to compute the *complete* power dispersion curve of nonlinear structures exhibiting the jump if computed by the conventional solution technique. With this new approach, the jump is successfully explained and a useful bistability phenomenon is revealed. It is further demonstrated how to modify the conventional solution technique for simulating the bistability phenomenon.

Among various nonlinear optical waveguides, planar structures have been extensively investigated with quasi-2D (using one transverse spatial variable) formulations. However, the quasi-2D formulations are expected to be valid only for weak nonlinear effects as nothing prevents a modal field from focussing or defocussing in the second transverse direction in the presence of strong nonlinear action. To verify the above prediction, a symmetric planar nonlinear structure is simulated in quasi-3D with a full vectorial formulation. The computed results are compared with those from the scalar quasi-2D simulation. Contrary to one's expectation, the self-focussing mechanism associated with weak nonlinear effects in a nonlinear planar structure is much more complicated than in a nonlinear channel structure and is very difficult to characterise. The proper modelling of these planar structures is fully discussed. That is the last example analysed in Chapter 6.

The accurate analysis of nonlinear optical waveguides with a full vectorial formulation is rather expensive. In Chapter 7, a scalar approximation procedure for weakly-guiding structures is established to ease the computation, which is also useful for qualitative analysis in the initial stages of design of structures other than weakly-guiding ones. The quantitative comparison of its results and those of the vectorial formulation is given for both planar and channel structures. Also, several precautions in utilising the scalar approximation in a nonlinear situation are discussed.

All the above is concerned with modal analysis of nonlinear optical waveguides. For those procedures to make sense, at least some of the predicted nonlinear modes must be stable during guided propagation. In other words, nonlinear guided waves can be investigated only by propagation methods if all nonlinear modes happen to be unstable. Therefore, the stability analysis of nonlinear modes is crucial to the application of nonlinear modal methods. Though the stability of nonlinear modes in quasi-2D structures has been widely reported, a nonlinear mode which is stable with one transverse-spatial-variable simulation may not be stable with two transverse-spatial-variable simulation in the presence of strong nonlinear effects. Therefore, there is an urgent need to perform the stability analysis of nonlinear modes in general quasi-3D nonlinear structures.

The numerical computation of the propagation of full vector waves in a three-dimensional (3D) nonlinear waveguide structure for several hundred steps is daunting and perhaps a supercomputer is needed to perform such an analysis. Fortunately, for many weakly-guiding structures, the scalar approximation yields almost identical results to those from the accurate vectorial formulation in the context of modal analyses, which allows us to perform the propagation analysis with nonlinear scalar waves rather than vector waves. The scalar nonlinear wave propagation in this book is achieved by the *finite-element plus finite-difference method* (FEM-FDM), which has several advantages over the conventional *beam propagation method* based on the *fast Fourier transform*. With this numerical procedure, the stability of nonlinear modes is investigated, and the bistability phenomenon predicted by the modal method is reexamined. That is also included in Chapter 7.

Linear and nonlinear coupled optical waveguides are most useful devices in optical signal processing. In analysing coupled optical waveguides three known techniques are frequently used, namely, superposition of supermodes or arraymodes, coupled-mode theories and propagation methods. The application of these three techniques to coupled waveguides is discussed in Chapter 8 in terms of their appropriateness, accuracy and computational efficiency. In particular, linear and nonlinear coupled-mode equations, applicable to vector-mode coupling in anisotropic and lossy media, are derived by using a reciprocity approach. Note that all these three techniques cry out for modal analyses. How to apply modal methods developed in the previous chapters to coupled waveguides is fully discussed in Chapter 8. Moreover, an example of coupled waveguides is analysed by the propagation method combined with the modal method in the linear case and in the nonlinear case.

The finite element method adopted in this book is a very powerful tool for linear and nonlinear optical waveguide computation. However, the finite-element programming requires expertise in both mathematics and physics/engineering plus great care and patience in order to produce accurate, efficient and user-friendly codes. Even if a finite-element program is available for use, the preparation of a suitable mesh for a complicated structure is usually a formidable task if done manually. In view of this, a powerful, self-contained and user-friendly software package ANOWS (Advanced Nonlinear Optical Waveguide Simulator), stored in a floppy diskette, is attached to this book for general use. This software package is devoted to the modal solution of linear/nonlinear and planar/channel optical waveguides. The formulation may be chosen from scalar or vector and from electric or magnetic fields. Automatic mesh generation programs are also included. Furthermore, the generated meshes and the computed modal fields can be plotted by GNUPLOT, which is widely available, by using the data conversion programs supplied. The ANOWS contents, documentation, usage and examples are presented in the Appendices.

1.3   Miscellaneous

- Owing to limited computation resources, the results presented in this book are estimated to be correct to about *four* significant figures. However, it is believed that no new features will be derived from more accurate computation.

- The number of symbols appearing in this book is enormous. All symbols used have a global span unless otherwise specified locally. To minimise the confusion that may arise from using so much notation, standard conventions in electromagnetics and mathematics are closely followed.

- To ensure ANOWS is robust and user-friendly, all of the computer programs were carefully modified since the results contained in this book were produced. Consequently, the number of iterations and the CPU time required for a solution may vary slightly from those presented in §6.3.

Chapter 2

Formulation of Electromagnetic Wave Equations for Nonlinear Optical Waveguides

2.1 Introduction

Nonlinear optical waveguides can be formulated in a linear way provided that an iteration scheme is incorporated to tackle the nonlinearity, as the nonlinear coefficients of known optical materials are very small [1]-[6]. The homogeneous electromagnetic wave equation for the solution of optical waveguides can be formulated either in a vectorial form [7] or in a scalar form as a weak-guidance approximation [8]. For most applications, optical waveguides should support only one or two modes, that is, they are weakly guiding in nature. Thus optical waveguides can often be formulated in a scalar form which requires less computation. The scalar formulation, however, cannot account for the polarisation effects. As a general approach, a vectorial formulation is preferred.

Among vectorial formulations the only one favoured by many is the H-vector version [9]-[12], which is particularly convenient for waveguides with permittivity discontinuities because continuity of the H-vector field components is automatically satisfied, as compared (i) with the E-vector version [13]-[15], which has difficulties in handling the discontinuity at dielectric interfaces, and (ii) with the full field formulation (both E and H)

11

[16], which not only has the discontinuity problem but requires more computation, and (iii) with the longitudinal (E_z, H_z) [17, 18] and transverse $(E_t$ or $H_t)$ formulations [19]-[22], which require restricted anisotropy. To simulate electrically-nonlinear optical effects, however, the H-vector formulation needs further computation and approximation to obtain the electric field during the iteration process [23]-[25] since the permittivities of nonlinear optical materials are now functions of the electric field. Another undesired effect with the H-vector formulation is that the modified nonlinear permittivity during the iteration process is discontinuous across inter-element interfaces whenever the FEM with C^0-elements is applied. This discontinuity might cause errors to accumulate during the iteration. Thus, a formulation valid for both the E-vector and the H-vector is desired so that their mutual consistency can be checked.

Although optical waveguides, generally speaking, have open boundaries, the closed-boundary approximation is adopted and would be sufficient for most applications in the area of guided optics with modal analysis since only guided modal fields are of interest. This truncation approach is more convenient than several other approaches:

- the method of dampers [26], which is inaccurate at the lowest order for two or three dimensional problems and becomes difficult to handle within a finite element model for any higher order of damper;

- the boundary integral and the analytical approach [20], [27]-[31], which not only has difficulties in treating the anisotropic and/or inhomogeneous media near the boundary but makes the final matrix complex, dense and unsymmetrical in general;

- the infinite element method [32]-[36], which involves the difficulty of finding proper decay parameters or mapping functions.

The efficiency of the truncation approach when using the FEM can be improved by using larger elements near the boundary. Of course, one has to admit that the truncation approach is very inefficient for calculating cut-off frequencies of an open structure. In such a case, the infinite element method may be preferred.

The constitutive relation between the electric field E and the electric flux density D is rather complicated in general nonlinear media, and depends on the detailed origin of the nonlinearity and the operating frequency [37, 38]. Nonlinear dielectric susceptibilities are related to the microscopic structure of the media and can be properly evaluated only with a full quantum-mechanical calculation. At the present stage of this work, the nonlinear materials considered are lossless and nondispersive, and the nonlinear effect is restricted to be self-focussing and/or self-defocussing [39]-[43].

The numerical solution of partial-differential equations in terms of the projection method (general method of moments or method of weighted residuals) [44] is more direct than in terms of the variational method in the sense that no variational functional needs to be found. The two methods are in fact are closely related [45, 46], both methods requiring two finite sets of basis functions, $\{e_n\}$ and $\{w_n\}$. One set, called the set of expansion functions, is used to approximate or expand the unknown field V, while the other, called the set of weighting or testing functions, is used to test the residual of the partial-differential operator equation (in the projection method) or expand the unknown adjoint field, V^a (in the variational method). Also, by using integration by parts, both methods can be expressed in "intermediate form" whereby the two finite sets of basis functions need to be piecewise differentiable only up to half the order of the original partial-differential operator (in this work, only piecewise differentiable to first order). Consequently, the variational method has no particular advantage over the general method of moments except that for some problems the functional of the variational method itself may represent a certain physical parameter, such as energy. The outcome is a weak formulation of the partial-differential equation that reduces to a matrix equation when the two finite sets of basis functions are substituted.

With the general method of moments, the basis functions are usually required to satisfy the associated discontinuity and boundary conditions for a physical solution. To relax such imposed restrictions, an extended operator is introduced.

It has been reported extensively that the E-vector and the H-vector formulations based on the conventional finite-element approach yield many spurious solutions [47]-[53] and these unphysical solutions can corrupt the whole modal spectrum [54, 55]. To suppress these spurious modes, a penalty term is added to the original formulation to enforce the solenoidal field constraint.

2.2   Nonlinear Wave Equations and Constitutive Relations

The waveguide structures considered are uniform along the propagation direction z and have arbitrary profiles in the xy-plane. The dielectric media under consideration are source-free and non-dissipative. The electromagnetic wave equation is therefore of homogeneous form and the permittivity and permeability dyadics are hermitian. When the dispersion of the media is relatively small and the spectrum of the laser beam is narrow enough, time-harmonic analysis applies. Due to the duality of the E-field and the H-field, the homogeneous and time-harmonic nonlinear wave equation has the same form for both the electric and the magnetic field. It can be written in the following general form with the phase factor $\exp\{j(\omega t - \beta z)\}$ being implied:

$$\nabla \times (\hat{p}^{-1} \cdot (\nabla \times V)) - k_0^2 \hat{q} \cdot V = 0 \qquad (k_0^2 \neq 0) \qquad (2.1)$$

defined in $\Omega \subset \boldsymbol{R}^2$ with homogeneous boundary and interface conditions:

$$\begin{cases} \boldsymbol{n} \times \boldsymbol{V} = \boldsymbol{0} & \text{on } C_1 \\ \boldsymbol{n} \times (\hat{p}^{-1} \cdot (\nabla \times \boldsymbol{V})) = \boldsymbol{0} & \text{on } C_2 \\ (\boldsymbol{n} \times \boldsymbol{V})_{diff} = \boldsymbol{0} & \text{on } \Sigma \\ (\boldsymbol{n} \times (\hat{p}^{-1} \cdot (\nabla \times \boldsymbol{V})))_{diff} = \boldsymbol{0} & \text{on } \Sigma \end{cases} \qquad (2.2)$$

where the subscript *diff* denotes the difference between quantities on the interior side and on the exterior side of Σ, and

$\nabla \quad \triangleq \quad \boldsymbol{a}_x \frac{\partial}{\partial x} + \boldsymbol{a}_y \frac{\partial}{\partial y} + \boldsymbol{a}_z \frac{\partial}{\partial z}$;

$\boldsymbol{R}^2 \quad - \quad$ the two-dimensional Euclidean space;

$\overline{\Omega} \quad - \quad$ a connected and closed set in $\boldsymbol{R}^2$;

$\Omega \quad - \quad \overline{\Omega}$ strictly excluding all boundaries and media interfaces — an open set;

$\omega \quad - \quad$ the angular frequency;

$\beta \quad - \quad$ the propagation constant;

$k_0 \quad \triangleq \quad \omega/c = \omega\sqrt{\mu_0 \epsilon_0}$, *i.e.* the free space wave number;

$C_1 \quad - \quad$ the Dirichlet boundary;

$C_2 \quad - \quad$ the Neumann boundary;

$\Sigma \quad - \quad$ the interface between different media;

$\boldsymbol{n} \quad - \quad$ the outward unit vector normal to Σ or $C_1 \cup C_2$.

The mapping of $\hat{p}$, $\hat{q}$ and V to the relative permeability $\hat{\mu}_r$, the relative permittivity $\hat{\epsilon}_r$, the electric field E and the magnetic field H is tabulated in Table 2.1. Here, $\hat{\epsilon}_r$ and $\hat{\mu}_r$ are restricted to be hermitian dyadics, which describe lossless optical materials.

Formulation	$\hat{p}$	$\hat{q}$	V
$\boldsymbol{E}$-vector	$\hat{\mu}_r$	$\hat{\epsilon}_r$	$\boldsymbol{E}$
$\boldsymbol{H}$-vector	$\hat{\epsilon}_r$	$\hat{\mu}_r$	$\boldsymbol{H}$

Table 2.1: The mapping of the variables for the two different formulations.

The nonlinear permittivity dyadic can be expressed as a linear term plus a nonlinear perturbation term:

$$\hat{\epsilon}_r = \hat{\epsilon}_r^l + \hat{\epsilon}_r^n \tag{2.3}$$

The nonlinear term $\hat{\epsilon}_r^n$ as a function of $\boldsymbol{E}$ is restricted to be monotonic and such that

$$\|\hat{\epsilon}_r^n\| \ll \|\hat{\epsilon}_r^l\| \quad \forall \boldsymbol{E} \tag{2.4}$$

so that a simple iteration scheme applies. Equation 2.4 also implies that $\hat{\epsilon}_r^n$ is saturable. The saturation condition of the nonlinearity is essential for a physical solution when the self-focussing action is stronger than the diffraction effect, as will be evident in Chapter 6.

2.3 Formulation of Electromagnetic Wave Equations and the Self-Adjointness of the Operator

Let L be the linear differential operator mapping a suitable linear space of functions $V(\Omega)$ into another linear space of functions $U(\Omega)$ defined by:

$$Lv \triangleq \overline{\nabla} \times (\hat{p}^{-1} \cdot (\overline{\nabla} \times \boldsymbol{v})) \qquad \forall \boldsymbol{v} \in V(\Omega), \tag{2.5}$$

with the associated discontinuity and boundary conditions defined in Equation 2.2. Let $V(\Omega)'$ and $U(\Omega)'$ be the duals of $V(\Omega)$ and $U(\Omega)$ for the *duality pairings* $< a, b >_V$ $(a \in V(\Omega)', \ b \in V(\Omega))$ and $< a, b >_U$ $(a \in U(\Omega)', \ b \in U(\Omega))$ [56]-[58], respectively. In this work, the duality pairing on $W(\Omega)' \times W(\Omega)$ for a linear space $W(\Omega)$ $(W \in \{U, V\})$ and its dual is defined by

$$< a, b >_W \triangleq \int_\Omega a^* \cdot b \ d\bar{x} d\bar{y}, \qquad a \in W(\Omega)', \quad b \in W(\Omega). \qquad (2.6)$$

where $*$ denotes complex conjugation. Here we have introduced a *normalisation factor* ρ, which can be β or k_0 or any other desired parameter having dimensions of m^{-1}, so that

$$\bar{\nabla} \triangleq \frac{1}{\rho} \nabla, \qquad \bar{x} \triangleq \rho x, \qquad \bar{y} \triangleq \rho y, \qquad \bar{z} \triangleq \rho z. \qquad (2.7)$$

Now, $\forall u \in U(\Omega)', \ v \in V(\Omega)$, we have the following weak formulation of the operator L:

$$
\begin{aligned}
< u, Lv >_U \ &\triangleq\ \int_\Omega u^* \cdot Lv \ d\bar{x} d\bar{y} \\
&=\ \int_\Omega u^* \cdot (\bar{\nabla} \times (\hat{p}^{-1} \cdot (\bar{\nabla} \times v))) \ d\bar{x} d\bar{y} \\
&=\ \int_\Omega (\bar{\nabla} \times u)^* \cdot \hat{p}^{-1} \cdot (\bar{\nabla} \times v) \ d\bar{x} d\bar{y} \\
&\quad -\oint_{\Sigma_-^+ + C} (n \times u)^* \cdot \hat{p}^{-1} \cdot (\bar{\nabla} \times v) \ d\bar{s} \\
&=\ \int_\Omega (\bar{\nabla} \times (\hat{p}^{\dagger-1} \cdot (\bar{\nabla} \times u)))^* \cdot v \ d\bar{x} d\bar{y} \\
&\quad + \Gamma(u, v) \\
&\triangleq\ < L'u, v >_V \ + \Gamma(u, v) \qquad (2.8)
\end{aligned}
$$

where the *bilinear concomitant* (or *conjunct*) [59, 60] $\Gamma(u, v)$, which contains boundary terms produced by the integrations by parts, is given by

$$
\begin{aligned}
\Gamma(u, v) \ \triangleq\ &-\oint_{\Sigma_-^+ + C} (n \times u)^* \cdot (\hat{p}^{-1} \cdot (\bar{\nabla} \times v)) \ d\bar{s} \\
&+ \oint_{\Sigma_-^+ + C} (\hat{p}^{\dagger-1} \cdot (\bar{\nabla} \times u))^* \cdot (n \times v) \ d\bar{s} \qquad (2.9)
\end{aligned}
$$

and $\Sigma_-^+ = \Sigma^+ \cup \Sigma_-$ denotes integration over both the internal (+) and the external (−) sides of the dielectric interface, $d\bar{s} = \rho \, ds$ the normalised line element, n the outward unit vector normal to the dielectric interface Σ or the boundary $C = C_1 \cup C_2$, † the hermitian conjugate and L' the transpose of L mapping $U(\Omega)'$ into $V(\Omega)'$.

Next consider the linear (multiplicative) operator M mapping $V(\Omega)$ into $U(\Omega)$ defined by

$$Mv \overset{\triangle}{=} \hat{q} \cdot v \qquad \forall v \in V(\Omega). \qquad (2.10)$$

Then we have

$$<u, Mv>_U \overset{\triangle}{=} \int_\Omega u^* \cdot \hat{q} \cdot v \, d\bar{x}d\bar{y}$$
$$= \int_\Omega (\hat{q}^\dagger \cdot u)^* \cdot v \, d\bar{x}d\bar{y}$$
$$\overset{\triangle}{=} <M'u, v>_V . \qquad (2.11)$$

To check the self-adjointness of the operators L and M, we consider now the spaces $V(\Omega)$ and $U(\Omega)$ as *pivot spaces* [56, 57, 61], that is, we identify $V(\Omega)$ and $U(\Omega)$ with their duals $V(\Omega)'$ and $U(\Omega)'$, respectively, and identify the duality pairings with the associated *inner products*. Then the pivot spaces L' and M' are called the *adjoints* of L and M, respectively, denoted by L^a and M^a. As $<\cdot, \cdot>_V$ and $<\cdot, \cdot>_U$ take exactly the same form, their subscripts will be omitted in the following, and it is assumed that any inner product without a subscript is interpreted in the space in which its arguments reside.

It is apparent that the "volume rules" of L^a and L are identical in Ω, and so are those of M^a and M, provided that $\hat{p}$ and $\hat{q}$ are hermitian dyadics. Thus, the differential operator L is *formally self-adjoint* [59], and M is *self-adjoint* [59, 62, 63] as it involves neither interface nor boundary terms.

Being formally self-adjoint, the operator L is self-adjoint *iff* the "surface rules" of L^a are the same as those of L given in Equation 2.2 [59]. The "surface rules" of L^a can be found as follows.

The integral terms along the dielectric interface in Equation 2.9 are of the form

$$\oint_{\Sigma_-^+} a \cdot b \, d\bar{s} = \oint_\Sigma (a_{diff} \cdot b_{av} + a_{av} \cdot b_{diff}) \, d\bar{s} \qquad (2.12)$$

where

$$\{\cdot\}_{diff} \overset{\triangle}{=} \{\cdot\}_{int} - \{\cdot\}_{ext}; \qquad \{\cdot\}_{av} \overset{\triangle}{=} \tfrac{1}{2}(\{\cdot\}_{int} + \{\cdot\}_{ext}) \qquad (2.13)$$

and Σ denotes just a single integration over the dielectric interface itself using an outward unit normal vector.

Now, Equation 2.9 can be written in the form of:

$$
\begin{aligned}
\Gamma(\boldsymbol{u}, \boldsymbol{v}) \;=\; & \oint_{\Sigma} [\boldsymbol{u}_{av}^{*} \cdot (\boldsymbol{n} \times (\hat{p}^{-1} \cdot (\overline{\nabla} \times \boldsymbol{v})))_{diff} \\
& + (\hat{p}^{\dagger-1} \cdot (\overline{\nabla} \times \boldsymbol{u}))_{av}^{*} \cdot (\boldsymbol{n} \times \boldsymbol{v})_{diff}] d\bar{s} \\
& - \oint_{\Sigma} [(\boldsymbol{n} \times (\hat{p}^{\dagger-1} \cdot (\overline{\nabla} \times \boldsymbol{u})))_{diff}^{*} \cdot \boldsymbol{v}_{av} \\
& + (\boldsymbol{n} \times \boldsymbol{u})_{diff}^{*} \cdot (\hat{p}^{-1} \cdot (\overline{\nabla} \times \boldsymbol{v}))_{av}] d\bar{s} \\
& + \int_{C_1} (\hat{p}^{\dagger-1} \cdot (\overline{\nabla} \times \boldsymbol{u}))^{*} \cdot (\boldsymbol{n} \times \boldsymbol{v}) d\bar{s} \\
& - \int_{C_1} (\boldsymbol{n} \times \boldsymbol{u})^{*} \cdot (\hat{p}^{-1} \cdot (\overline{\nabla} \times \boldsymbol{v})) \, d\bar{s} \\
& + \int_{C_2} \boldsymbol{u}^{*} \cdot (\boldsymbol{n} \times (\hat{p}^{-1} \cdot (\overline{\nabla} \times \boldsymbol{v}))) \, d\bar{s} \\
& - \int_{C_2} (\boldsymbol{n} \times (\hat{p}^{\dagger-1} \cdot (\overline{\nabla} \times \boldsymbol{u})))^{*} \cdot \boldsymbol{v} \, d\bar{s}. \qquad (2.14)
\end{aligned}
$$

The "surface rule" of L^a imposed on $\boldsymbol{u}$ requires that the bilinear conjunct vanishes [59]. Hence, by inspecting Equation 2.14, we obtain

$$
\begin{cases}
\boldsymbol{n} \times \boldsymbol{u} = \boldsymbol{0} & \text{on } C_1 \\
\boldsymbol{n} \times (\hat{p}^{\dagger-1} \cdot (\nabla \times \boldsymbol{u})) = \boldsymbol{0} & \text{on } C_2 \\
(\boldsymbol{n} \times \boldsymbol{u})_{diff} = \boldsymbol{0} & \text{on } \Sigma \\
(\boldsymbol{n} \times (\hat{p}^{\dagger-1} \cdot (\nabla \times \boldsymbol{u})))_{diff} = \boldsymbol{0} & \text{on } \Sigma
\end{cases}
\qquad (2.15)
$$

which are the same as those of L in Equation 2.2 for hermitian dyadics $\hat{p}$ and $\hat{q}$. Thus, the differential operator L, with the associated discontinuity and boundary conditions in Equation 2.2, is self-adjoint. The self-adjointness of L and M has a great bearing on Equation 2.1, as is evident in the next section.

2.4 Eigenvalue Problems and Moment Methods

2.4.1 Eigenvalue problems

Our "linear" eigenvalue problem (Equation 2.1) can be written in the form of the linear operator equation

$$Lv \;=\; \lambda Mv \qquad\qquad (2.16)$$

with the associated discontinuity and boundary conditions given in Equation 2.2. Here $\lambda = (k_0/\rho)^2$. Note that $\lambda \equiv 1$ when the normalisation factor $\rho = k_0$. In such a situation, λ is formally taken as an eigenvalue and the detailed solution technique will be described in Chapter 5.

Although operators L and M have their own domains, D_L and D_M, for our purposes we define the abstract inner-product space $D_L \cap D_M$ so that Equation 2.16 makes sense for any $v \in D_L \cap D_M$; one may call it the domain of L and M. Then we have the following theorem:

Theorem 2.1 *All eigenvalues of Equation 2.16 are real if M is positive definite and both L and M are self-adjoint in $D_L \cap D_M$.*

Proof: Let (λ, v) be an eigenpair of Equation 2.16. One then has

$$
\begin{aligned}
0 &= <v, Lv> - <Lv, v> \\
&= <v, \lambda Mv> - <\lambda Mv, v> \\
&= (\lambda - \lambda^*) <v, Mv>
\end{aligned}
$$

which implies that $\lambda \equiv \lambda^*$.   #

As is shown in the previous section, both L and M are self-adjoint for hermitian dyadics $\hat{p}$ and $\hat{q}$. Furthermore, it can be easily shown that M is also positive definite provided that $\hat{q}$ is a hermitian dyadic in non-plasma media. Therefore, the exact eigenvalues resulting from Equation 2.16 form a real set.

2.4.2 Method of moments

Given Equation 2.16, we choose a finite set of *expansion* functions $v_n \in D_L \cap D_M$, $n = 1, 2, 3, \cdots$, and let the approximate solution be expanded as

$$v = \sum_n \alpha_n v_n \tag{2.17}$$

where the α_n are unknown constant coefficients. We then choose a finite set of *weighting* (or *testing*) functions $w_m \in D_{L^a} \cap D_{M^a}$, $m = 1, 2, 3, \cdots$. Taking the inner-product of Equation 2.16 with each w_m after substituting Equation 2.17, we obtain

$$\sum_n < w_m, Lv_n > \alpha_n = \lambda \sum_n < w_m, Mv_n > \alpha_n \quad \forall m = 1, 2, 3, \cdots$$
$$\tag{2.18}$$

where $< w_m, Lv_n >$ and $< w_m, Mv_n >$ take exactly the same forms as in Equations 2.8 and 2.11 with u and v being replaced by w_m and v_n, respectively.

Equation 2.18 can be written in a matrix form

$$A\alpha \;=\; \lambda B\alpha \tag{2.19}$$

by the defining the matrix elements

$$A_{mn} \;\equiv\; < w_m, Lv_n > \tag{2.20}$$
$$B_{mn} \;\equiv\; < w_m, Mv_n > \tag{2.21}$$

and the column vector

$$\alpha \;\equiv\; [\alpha_1, \alpha_2, \alpha_3, \cdots]^T \tag{2.22}$$

where $m, n = 1, 2, 3, \cdots$.

Equation 2.19 is a generalised algebraic eigenvalue matrix equation. Its eigenvalues may not be real since they are only approximations to the exact eigenvalues of Equation 2.16. If each $w_m = v_m$ ($m = 1, 2, 3, \cdots$), however, the method of moments is known as Galerkin's method, and then the matrices A and B are hermitian since L and M are self-adjoint. The hermitian eigenproblem (Equation 2.19) then has real eigenvalues and can be solved by the techniques described in Chapter 5. The computation of real eigenvalues and eigenvectors is much less expensive than the computation of complex ones.

2.5   Extended Operator

The expansion and testing functions discussed in the previous section are constrained to satisfy the discontinuity and boundary conditions in

Equations 2.2 and 2.15 since they must be contained in D_L and D_{L^a}, respectively. To relax such restrictions while retaining self-adjointness, we introduce an extended operator L_e in the following. The method has been successfully employed by Cambrell [45, 65], Harrington [46], Friedman [59], Lanczos [64], Cambrell and Williams [66], and others.

The extended operator procedure becomes particularly important in multidimensional problems as it is not always easy to find simple expansion functions in the domain of the original operator. With the operator extended, a wider class of basis functions can be used for solution by the method of moments [46].

Following Cambrell's technique [45, 65], the differential operator L defined in Equation 2.5 can be extended by extending its domain $D_L(\Omega)$ to the domain $D_{L_e}(\bar{\Omega})$ (consequently, $D_{L^a}(\Omega)$ to $D_{L_e^a}(\bar{\Omega})$). The extended operator, which combines both the "volume rule" and the "surface rule" into one entity, is given by

$$
\begin{aligned}
< u, L_e v >_e \;\stackrel{\triangle}{=}\; & < u, Lv > \\
& - \oint_\Sigma [u_{av}^* \cdot (n \times (\hat{p}^{-1} \cdot (\overline{\nabla} \times v)))_{diff} \\
& + (\hat{p}^{\dagger -1} \cdot (\overline{\nabla} \times u))_{av}^* \cdot (n \times v)_{diff}] d\bar{s} \\
& - \int_{C_1} (\hat{p}^{\dagger -1} \cdot (\overline{\nabla} \times u))^* \cdot (n \times v) d\bar{s} \\
& - \int_{C_2} u^* \cdot (n \times (\hat{p}^{-1} \cdot (\overline{\nabla} \times v))) \, d\bar{s} \qquad (2.23) \\
= & \int_\Omega (\overline{\nabla} \times u)^* \cdot \hat{p}^{-1} \cdot (\overline{\nabla} \times v) d\bar{x} d\bar{y} \\
& - \oint_\Sigma [(\hat{p}^{\dagger -1} \cdot (\overline{\nabla} \times u))_{av}^* \cdot (n \times v)_{diff} \\
& + (n \times u)_{diff}^* \cdot (\hat{p}^{-1} \cdot (\overline{\nabla} \times v))_{av}] \, d\bar{s} \\
& - \int_{C_1} [(\hat{p}^{\dagger -1} \cdot (\overline{\nabla} \times u))^* \cdot (n \times v) \\
& + (n \times u)^* \cdot (\hat{p}^{-1} \cdot (\overline{\nabla} \times v))] \, d\bar{s} \qquad (2.24) \\
= & < L^a u, v > \\
& - \oint_\Sigma [((n \times (\hat{p}^{\dagger -1} \cdot (\overline{\nabla} \times u))))_{diff}^* \cdot v_{av} \\
& + (n \times u)_{diff}^* \cdot (\hat{p}^{-1} \cdot (\overline{\nabla} \times v))_{av}] \, d\bar{s}
\end{aligned}
$$

$$-\int_{C_1} (\boldsymbol{n} \times \boldsymbol{u})^* \cdot (\hat{p}^{-1} \cdot (\overline{\nabla} \times \boldsymbol{v})) \, d\bar{s}$$

$$-\int_{C_2} (\boldsymbol{n} \times (\hat{p}^{\dagger-1} \cdot (\overline{\nabla} \times \boldsymbol{u})))^* \cdot \boldsymbol{v} \, d\bar{s} \qquad (2.25)$$

$$\stackrel{\triangle}{=} \; < L_e^a \boldsymbol{u}, \boldsymbol{v} >_e \qquad\qquad (2.26)$$

where $< \cdot, \cdot >_e$ is an extension of the inner product $< \cdot, \cdot >$ defined in Equation 2.6. Equations 2.23 to 2.25 are called the *original form*, the *intermediate form* and the *adjoint form*, respectively, of the weak formulation of the extended operator, and these names will be used in a later chapter. (Actually (2.23)-(2.25) is just a rearrangement of Equation 2.8.)

It can be shown that the "rules" of L_e^a and L_e are identical in $\bar{\Omega} = \Omega \cup \Sigma \cup C_1 \cup C_2$. Hence the extended operator L_e is also self-adjoint in spite of the discontinuity and boundary conditions (Equation 2.2).

Actually the operator M can be conceptually extended to include all the given source terms of the problem, but since the problem is homogeneous the given source terms are all zero.

In summary, the weak formulation of the differential equation (Equation 2.1) is thus

$$< \boldsymbol{u}, L_e \boldsymbol{v} >_e \; = \; \lambda < \boldsymbol{u}, M\boldsymbol{v} >_e \qquad\qquad (2.27)$$

in the extended sense. A projection solution proceeds in this extended domain in the same manner as in the original domain (with extended source terms if the differential equation is inhomogeneous), but the expansion and testing functions need not satisfy all of the discontinuity and boundary conditions (Equation 2.2). For best numerical results, however, it is preferable to satisfy *a priori* as many of the required interface and boundary conditions as possible.

2.6   Spurious Mode Problem and Penalty Function Method

Exact solutions to the differential-operator equation (Equation 2.1) must satisfy the solenoidal constraint

$$\nabla \cdot (\hat{q} \cdot \boldsymbol{V}) = 0 \qquad \text{in } \Omega \qquad\qquad (2.28)$$

unless $k_0^2 = 0$, corresponding to static fields which are of no interest here. In a finite element model, if the spaces of *admissible functions* (both expansion and testing functions) are not restricted by the solenoidal constraint, many spurious solutions appear [67] and these spurious solutions can corrupt the whole spectrum. Even if Equation 2.28 is satisfied by admissible functions, some undesired solutions with zero eigenvalues can still exist as the solenoidal subspace itself may include a non-trivial irrotational sub-subspace. It is desired that admissible functions are defined in a solenoidal subspace with non-vanishing curl except for the trivial *zero* vector function. However, constructing such a subspace for waveguide problems has proven to be very difficult even before considering the discontinuity and boundary conditions.

It should be mentioned that admissible functions based on edge-elements [68, 69] are solenoidal and have a constant curl in each element for first-order edge-elements. They have been applied to 3D eddy-current problems [70]-[74], scattering problems [75] and 3D cavities [76] successfully. However, there are difficulties in generalising the method to waveguide structures for solution of guided modes as there are no edges in the propagation direction for representing the longitudinal component of the field, let alone knowing how to incorporate the z-dependent factor $\exp\{-j\beta z\}$ into the admissible functions based on edge-elements. Although a modified version of edge-elements called *tangential elements* [50] for the solution of guided modes is available, where the transverse components of the field are approximated by edge-elements and the longitudinal component is approximated by Lagrangian finite elements[1], it seems that the results from these elements are not as accurate as those from other methods with very similar computations apart from the fact that these elements produce highly degenerate unphysical modes of zero eigenvalues. The use of higher-order tangential elements has been reported [51]. Their accuracy and efficiency as well as their applicability to anisotropic waveguides need further investigation. The above tangential elements fall into the category described below.

The restrictions on the spaces of admissible functions can be relaxed by including an irrotational subspace or properly accounting for the infinite-dimensional null space of the curl operator, combined with a proper choice of the solution method by which the non-zero eigenpairs can be found directly without bothering with those having zero eigenvalues, such as the *method of bisection and inverse iteration with shift* [77]. The finite elements suggested in [50, 51] and [78]-[80] fall into this

---

[1]Lagrangian finite elements are derived from Lagrangian interpolation, where the coefficients of interpolation involve only the field components.

category. These elements have not been widely accepted possibly due to inefficiency, or restricted applications or being too complicated to be implemented.

As it is difficult to construct satisfactory spaces of admissible functions for waveguide problems, an alternative approach is to solve Equation 2.1 with Equation 2.28 as a constraint while the expansion and testing functions need not satisfy the solenoidal condition. One might immediately think of the Lagrangian Multiplier Method or the Reduction Method [81, 83], but those methods are very inefficient.

A reduction method of constraint using Hermitian finite elements[2] was reported for inhomogeneous waveguides [84] where the reduction (elimination of the axial field component using the solenoidal constraint) is performed on an element basis and consequently the sparsity of the matrices in the global matrix equation is retained. It was claimed by its authors that the method is advantageous over the Lagrangian FEM as it requires fewer degrees of freedom than the latter to achieve the same accuracy. However, the efficiency of a method cannot be simply determined by the number of degrees of freedom it requires to achieve a certain accuracy as, in general, the bandwidth of the matrices resulting from Hermitian finite elements is larger than that resulting from Lagrangian ones with the same number of degrees of freedom; unfortunately the example the authors used for comparison is one where the sparsity of the matrices is destroyed. Therefore, further information is needed to justify the efficiency of the method.

The widely adopted method of constraint for waveguide structures is the penalty function method [14, 15, 48, 54], [85]-[89] though it may be a "stranger" to mathematicians. With the penalty function method the spurious modes will be shifted out of the range of interest if the penalty parameter is chosen properly. The difficulty involved in applying the method is the choice of the penalty parameter and the associated accuracy for higher-order modes. Fortunately, only one or two lowest-order modes are required for optical waveguide devices. Thus, in this work, the penalty function method can be successfully adopted to deflect the spurious modes.

---

[2]Hermitian finite elements are derived from Hermitian interpolation, where the coefficients of interpolation involve both the field components and their first-order derivatives.

The modified formulation with a penalty term added is given by:

$$
\begin{aligned}
< \boldsymbol{u}, \hat{L}_e \boldsymbol{v} >_e \; &\overset{\triangle}{=} \; < \boldsymbol{u}, L_e \boldsymbol{v} >_e + p < \boldsymbol{u}, L_p \boldsymbol{v} >_e \\
&= \; \lambda < \boldsymbol{u}, M \boldsymbol{v} >_e
\end{aligned}
\tag{2.29}
$$

with p being a positive penalty parameter and

$$
\begin{aligned}
< \boldsymbol{u}, L_p \boldsymbol{v} >_e \; \overset{\triangle}{=} \; &- \int_\Omega \boldsymbol{u}^* \cdot \hat{q} \cdot \overline{\nabla}\overline{\nabla} \cdot (\hat{q} \cdot \boldsymbol{v}) \, d\bar{x}d\bar{y} \\
&+ \oint_\Sigma [(\boldsymbol{n} \cdot \hat{q}^\dagger \cdot \boldsymbol{u})^*_{av}(\overline{\nabla} \cdot (\hat{q} \cdot \boldsymbol{v}))_{diff} \\
&- ((\overline{\nabla} \cdot (\hat{q}^\dagger \cdot \boldsymbol{u}))^*_{av}(\boldsymbol{n} \cdot \hat{q} \cdot \boldsymbol{v})_{diff}] \, d\bar{s} \\
&+ \int_{C_1} (\boldsymbol{n} \cdot \hat{q}^\dagger \cdot \boldsymbol{u})^* \overline{\nabla} \cdot (\hat{q} \cdot \boldsymbol{v}) \, d\bar{s} \\
&- \int_{C_2} (\overline{\nabla} \cdot (\hat{q}^\dagger \cdot \boldsymbol{u}))^*(\boldsymbol{n} \cdot \hat{q} \cdot \boldsymbol{v}) \, d\bar{s}
\end{aligned}
\tag{2.30}
$$

$$
\begin{aligned}
= \; &\int_\Omega [\overline{\nabla} \cdot (\hat{q}^\dagger \cdot \boldsymbol{u})]^*[\overline{\nabla} \cdot (\hat{q} \cdot \boldsymbol{v})] \, d\bar{x}d\bar{y} \\
&- \oint_\Sigma [(\overline{\nabla} \cdot (\hat{q}^\dagger \cdot \boldsymbol{u}))^*_{av}(\boldsymbol{n} \cdot \hat{q} \cdot \boldsymbol{v})_{diff} \\
&+ (\boldsymbol{n} \cdot \hat{q}^\dagger \cdot \boldsymbol{u})^*_{diff}(\overline{\nabla} \cdot (\hat{q} \cdot \boldsymbol{v}))_{av}] \, d\bar{s} \\
&- \int_{C_2} [(\overline{\nabla} \cdot (\hat{q}^\dagger \cdot \boldsymbol{u}))^*(\boldsymbol{n} \cdot \hat{q} \cdot \boldsymbol{v}) \\
&+ (\boldsymbol{n} \cdot \hat{q}^\dagger \cdot \boldsymbol{u})^* \overline{\nabla} \cdot (\hat{q} \cdot \boldsymbol{v})] \, d\bar{s}
\end{aligned}
\tag{2.31}
$$

$$
\begin{aligned}
= \; &- \int_\Omega (\hat{q}^\dagger \cdot \overline{\nabla}\overline{\nabla} \cdot (\hat{q}^\dagger \cdot \boldsymbol{u}))^* \cdot \boldsymbol{v} \, d\bar{x}d\bar{y} \\
&+ \oint_\Sigma [(\overline{\nabla} \cdot (\hat{q}^\dagger \cdot \boldsymbol{u}))^*_{diff}(\boldsymbol{n} \cdot \hat{q} \cdot \boldsymbol{v})_{av} \\
&- (\boldsymbol{n} \cdot \hat{q}^\dagger \cdot \boldsymbol{u})^*_{diff}(\overline{\nabla} \cdot (\hat{q} \cdot \boldsymbol{v}))_{av}] \, d\bar{s} \\
&+ \int_{C_1} (\overline{\nabla} \cdot (\hat{q}^\dagger \cdot \boldsymbol{u}))^*(\boldsymbol{n} \cdot \hat{q} \cdot \boldsymbol{v}) \, d\bar{s} \\
&- \int_{C_2} (\boldsymbol{n} \cdot \hat{q}^\dagger \cdot \boldsymbol{u})^* \overline{\nabla} \cdot (\hat{q} \cdot \boldsymbol{v}) \, d\bar{s}.
\end{aligned}
\tag{2.32}
$$

Here the weak formulation of the penalty operator L_p also has three forms: the *original* (Equation 2.30), the *intermediate* (Equation 2.31) and the *adjoint* (Equation 2.32). The line integral terms are included to relax the new restrictions [65, 88] imposed on expansion and testing functions at dielectric interfaces and boundaries due to the penalty term:

$$\begin{cases} \nabla \cdot (\hat{q} \cdot \boldsymbol{V}) = 0 & \text{on } C_1 \\ \boldsymbol{n} \cdot \hat{q} \cdot \boldsymbol{V} = 0 & \text{on } C_2 \\ (\nabla \cdot (\hat{q} \cdot \boldsymbol{V}))_{diff} = 0 & \text{on } \Sigma \\ (\boldsymbol{n} \cdot \hat{q} \cdot \boldsymbol{V})_{diff} = 0 & \text{on } \Sigma \end{cases} \tag{2.33}$$

and $\hat{L}_e$ is also self-adjoint in the extended sense. A moment solution can be obtained in the same way as in Equation 2.18 with L being replaced by $\hat{L}_e$.

It is worth noting that all line integral terms in Equations 2.23-2.25 and 2.30-2.32 will vanish if the expansion and testing functions satisfy all of the discontinuity and boundary conditions in Equations 2.2 and 2.33.

2.7   Guided Power

The guided power P is one of the most important parameters in non-linear optics. With k_0 and the dimensions of the structure given, the key dispersion relation in the analysis and design of nonlinear optical waveguides is the power dispersion curve of β vs P. The power P can be found by integrating the z-component of the Poynting vector over the cross-section Ω, giving

$$P = \frac{1}{2\bar{Z} k_0 \rho} \cdot \Im\{-\int_\Omega \boldsymbol{V}^* \times (\hat{p}^{-1} \cdot (\overline{\nabla} \times \boldsymbol{V})) \cdot \boldsymbol{a}_z \, d\bar{x} d\bar{y}\} \tag{2.34}$$

where $\Im\{\cdot\}$ denotes the imaginary part of the argument, and

$$\bar{Z} = \begin{cases} Z_0 & \text{for } \boldsymbol{V} \equiv \boldsymbol{E} \\ Z_0^{-1} & \text{for } \boldsymbol{V} \equiv \boldsymbol{H} \end{cases} \quad ; \quad \hat{p} = \begin{cases} \hat{\mu}_r & \text{for } \boldsymbol{V} \equiv \boldsymbol{E} \\ \hat{\epsilon}_r & \text{for } \boldsymbol{V} \equiv \boldsymbol{H} \end{cases} \tag{2.35}$$

with $Z_0 \triangleq \sqrt{\mu_0/\epsilon_0}$ being the intrinsic impedance in vacuum.

Chapter 3

Vector Finite Element Approach

3.1   Introduction

Many problems in physics and engineering are governed by (linear or nonlinear) partial-differential equations. These equations, especially in multivariable problems, are extremely difficult to solve directly even with some reasonable assumptions and approximations except for very simple problems of regular geometry with the most simple boundary conditions, even if the mathematical formulation is completed. The *finite element method* (FEM) or *finite element analysis* (FEA) [90], a technique of converting (ordinary or partial) differential equations into algebraic ones, provides a most powerful tool for solving those problems. The FEM produces the two sets of basis functions required by the projection method or the variational method in order to form a matrix equation.

The fundamental idea of the FEM is to approximate the dependent variables (*e.g.* the components of a vector field) of a practical problem locally in terms of simple functions defined over small, but not infinitesimal, elements. Once this is done, an algebraic matrix equation is produced through a projection or variational procedure. With the conventional vector finite element methods (FEMs), each dependent variable for a vectorial problem is approximated independently [91]. Consequently, the resulting basis functions are not admissible to problems with inherent constraints between dependent variables, and those constrained problems have to be solved by Lagrange multiplier or by penalty FEMs [57, 60, 92, 93]. Now there is a thrust towards developing non-conventional vector finite elements, in order to approximate vectorial problems with constraints [68, 73, 79, 80], [94]-[98], that takes the whole vector of dependent variables as one entity in the approximation. The advantage of such an approach is obvious as it may be possible to make trial bases satisfying part or all of the problem's constraints *a priori*. The numerical computations should be more efficient, reliable and accurate when trial bases satisfy as many of the constraints as possible.

In the area of electromagnetics, it is preferred that the trial bases satisfy the relevant solenoidal constraint, which can be done easily now for purely 2D or 3D problems, such as cavities and eddy currents, by using *edge-elements* [69, 94, 95]. Unfortunately, any attempt to make those edge-elements suitable for this work has proved to be disappointing due to its quasi-3D nature and the special form of the phase factor, as stated in the previous chapter, and no desirable alternatives have yet been found.

In the following it is intended to explore a systematic approach for constructing nonconventional vector finite elements so as to develop a robust procedure for solving quasi-3D waveguiding problems by using generalised vector finite elements. This approach provides a guideline for the construction of new vector finite elements for future problems and can naturally accommodate any forthcoming non-conventional vector finite elements into this work. In fact the form of elements finally employed in the present work is equivalent to the conventional one although they are actually constructed from the viewpoint of generalised vector finite elements.

The weak vectorial formulation of the homogeneous electromagnetic wave equations in the previous chapter, as discussed therein, is appropriate for the general method of moments, where the expansion and testing spaces are not specified in any particular way. In this chapter the expansion and testing spaces are constructed in terms of generalised vector finite elements. A matrix equation representing the original problem is then assembled by using *Galerkin's technique* whereby the testing space is chosen to be the same as the expansion space, that is, the two finite sets of vector basis functions are chosen to be identical. Also, it is shown how to construct special vector basis functions for enforcing interface and boundary conditions in media of arbitrary anisotropy.

3.2 Nonconventional Vector Finite Elements

With the conventional finite element method based on nodal variables, an approximation $\tilde{u}$ to a scalar function u within an element by Lagrangian interpolation is given by [91]

$$\tilde{u}^e(\bar{\boldsymbol{x}}) = \sum_{i=1}^{s} \bar{u}_i^e N_i^e(\bar{\boldsymbol{x}}) \tag{3.1}$$

where $\bar{\boldsymbol{x}}$ is a point in Ω^e, which is a subspace of an n-dimensional Euclidean space $\boldsymbol{R}^n$, given by

$$\bar{\boldsymbol{x}} = \sum_{j=1}^{n} \bar{x}_j \boldsymbol{a}_j \tag{3.2}$$

and where $\boldsymbol{a}_j$ $(j = 1, 2, \cdots, n)$ are the unit vectors of the global Cartesian coordinates, the superscript e denotes the element index, s is the number of nodes of the element e, and $\bar{u}_i^e$ and N_i^e $(i = 1, 2, \cdots, s)$ are nodal unknowns and *scalar shape functions* of the element e, respectively.

For first-order simplex elements in n-dimensions,

$$N_i^e = \xi_i \quad \forall i \in \sigma_s \tag{3.3}$$

where ξ_i $(\forall i \in \sigma_s)$ are local *area coordinates*[1] and σ_s denotes the set $\{1, 2, \cdots, s\}$. To retain the independence of the nodal values, s must be equal to $n+1$ since $\sum_{i=1}^{n+1} \xi_i = 1$. The transformation between the global and the local coordinates in element e is given by [91]

$$\bar{x}_j = \sum_{i=1}^{n+1} \bar{x}_{ji}^e \xi_i \qquad \forall j \in \sigma_n \tag{3.4}$$

where the subscripts i and j are the local and the global indices, respectively.

In the conventional vector FEM, each component of the vector field is approximated in the same way as in the scalar case and by Lagrangian interpolation each of the m dependent variables is given by [91]

$$\tilde{u}_k^e(\bar{\boldsymbol{x}}) = \sum_{i=1}^{s} \bar{u}_{k,i}^e N_i^e(\bar{\boldsymbol{x}}) \qquad \forall k \in \sigma_m \tag{3.5}$$

---

[1] They are sometimes called *natural* or *barycentric coordinates* in cases of other than two-dimensional simplexes.

where the subscript k is the field-component index. Hence these m dependent variables are approximated independently. Consequently, any inherent relations between these dependent variables, such as the solenoidal condition in electromagnetics and the incompressibility constraint in fluid mechanics, are ignored, which may lead to spurious solutions as well as computational inefficiency. It is highly desirable to incorporate such constraints into the approximation.

The approximation of a vector field described in [99, pp. 46-48] is more general conceptually in the sense that the whole vector is taken as one entity in the approximation:

$$\tilde{u}^e(\bar{x}) = \sum_{i=1}^{s} \bar{u}_i^e N_i^e(\bar{x}) \qquad \bar{x} \in \Omega^e \tag{3.6}$$

where $\bar{u}_i^e$ ($\forall i \in \sigma_s$) are the vector nodal unknowns defined by

$$\bar{u}_i^e = \sum_{k=1}^{m} a_k \bar{u}_{k,i}^e \tag{3.7}$$

and $N_i^e(\bar{x})$ ($\forall i \in \sigma_s$) are the scalar shape functions as before. Unfortunately, this procedure has no substantial advantage over the conventional one due to the shape functions employed being scalar and the same for all the components of the vector field. In the following we describe a potentially more useful procedure for the approximation of a vector field.

Instead of using vector unknowns and scalar shape functions, we use scalar unknowns α_r^e and *vector shape functions* $\psi_r^e(\bar{x})$ ($r \in \sigma_t$). Also, the unknowns are not necessarily the nodal values and could be those along edges of elements [69, 95] or simply abstract coefficients. The generalised approximation of a vector field within an element now reads:

$$\tilde{u}^e(\bar{x}) = \sum_{r=1}^{t} \alpha_r^e \psi_r^e(\bar{x}) \qquad \bar{x} \in \Omega^e \tag{3.8}$$

The finite elements on which the above generalised approximation is defined might as well be called *generalised vector finite elements* with the conventional case being a particular member of this family, where t is the total number of degrees of freedom per element and $t = m \times s$ for the conventional case with nodal unknowns.

It is apparent that any constraints between the m dependent variables can be enforced individually on each vector shape function $\psi_r^e(\bar{x})$

($\forall r \in \sigma_t$) as long as the usual convergence requirements [91, 100, 101] are not violated. Furthermore, when constructing special vector finite elements satisfying some inherent constraints, difficulties in satisfying certain continuity requirements across interelement interfaces are expected. Fortunately, the interelement continuity requirements can be relaxed by combining *hybrid FEMs* [60, 92] whenever that is an obstacle to employing the new vector finite elements.

In the following, m will be set equal to 3 and n equal to 2 since quasi-3D problems are to be solved.

3.3 Basis Functions and Element Matrix Equations

First, we introduce some notation:

$$L^2(\Omega) \quad - \quad \text{the usual Hilbert space of square integrable functions in the Lebesgue sense defined on } \Omega;$$

$$H^k(\Omega) \quad = \quad \{\phi \mid \tfrac{\partial^r \phi}{\partial \bar{x}^i \partial \bar{y}^j} \in L^2(\Omega), i, j \geq 0, i+j = r, r \leq k\},$$
a Sobolev space of order k;

$$(H^k(\Omega))^m \quad = \quad \{\boldsymbol{u} = \sum_{i=1}^m u_i \boldsymbol{a}_i \mid u_i \in H^k(\Omega), \forall i \in \sigma_m\};$$

$$C^k(\Omega) \quad = \quad \{\phi \mid \tfrac{\partial^r \phi}{\partial \bar{x}^i \partial \bar{y}^j} \text{ is continuous in } \Omega, i, j \geq 0,$$
$$i+j = r, r \leq k\}, \text{ a linear space of functions;}$$

C^k-element $\quad - \quad$ the type of elements on which the shape functions defined belong to $C^k(\Omega)$.

Next, the following letters have the following reserved meaning throughout this section when they are used as indices, parameters or arguments of functions:

e $\quad - \quad$ the element index; $e = 1, 2, \cdots, n_e$

g $\quad - \quad$ the global node index; $g = 1, 2, \cdots, n_s$

i $\quad - \quad$ the global index of basis functions; $i = 1, 2, \cdots, q$

k $\quad - \quad$ the field-component index; $k = 1, 2, 3$

l $\quad - \quad$ the local vertex-node index of the element e; $l = 1, 2, 3$

n_e $\quad - \quad$ the total number of elements;

n_s $\quad - \quad$ the total number of global nodes in the model;

q $\quad - \quad$ the total number of degrees of freedom; $q = 3n_s$

r $\quad - \quad$ the local index of basis functions; $r = 1, 2, \cdots, t$

t —   the number of degrees of freedom per element;
$t = 9$ for first-order elements, and
$t = 18$ for second-order elements.

The weak formulation of the homogeneous electromagnetic wave equations outlined in Chapter 2 has three forms: original, intermediate and adjoint. Any of the three forms can be adopted in FEA; however, only the intermediate form is generally employed because it requires the least smoothness imposed on the type of elements to be selected, as is evident from Table 3.1.

Form	Testing Space	Expansion Space	Element Type
Original	$(H^0(\Omega))^3$	$(H^2(\Omega))^3$	C^1
Intermediate	$(H^1(\Omega))^3$	$(H^1(\Omega))^3$	C^0
Adjoint	$(H^2(\Omega))^3$	$(H^0(\Omega))^3$	C^1

Table 3.1: A comparison of the three forms of the weak formulation.

Summarising Equations 2.24, 2.29 and 2.31, the intermediate form of the extended operator is given by:

$$
\begin{aligned}
< \boldsymbol{u}, \hat{L}_e \boldsymbol{v} >_e \; \triangleq \; & \int_\Omega (\overline{\nabla} \times \boldsymbol{u})^* \cdot \hat{p}^{-1} \cdot (\overline{\nabla} \times \boldsymbol{v}) d\bar{x} d\bar{y} - \int_\Omega \boldsymbol{u}^* \cdot \hat{q} \cdot \boldsymbol{v}\, d\bar{x} d\bar{y} \\
& - \oint_\Sigma [(\hat{p}^{\dagger -1} \cdot (\overline{\nabla} \times \boldsymbol{u}))_{av}^* \cdot (\boldsymbol{n} \times \boldsymbol{v})_{diff} \\
& + (\boldsymbol{n} \times \boldsymbol{u})_{diff}^* \cdot (\hat{p}^{-1} \cdot (\overline{\nabla} \times \boldsymbol{v}))_{av}]\, d\bar{s} \\
& - \int_{C_1} [(\hat{p}^{\dagger -1} \cdot (\overline{\nabla} \times \boldsymbol{u}))^* \cdot (\boldsymbol{n} \times \boldsymbol{v}) \\
& + (\boldsymbol{n} \times \boldsymbol{u})^* \cdot (\hat{p}^{-1} \cdot (\overline{\nabla} \times \boldsymbol{v}))]\, d\bar{s} \\
& + p\{ \int_\Omega (\overline{\nabla} \cdot (\hat{q}^\dagger \cdot \boldsymbol{u}))^* (\overline{\nabla} \cdot (\hat{q} \cdot \boldsymbol{v}))\, d\bar{x} d\bar{y} \\
& - \oint_\Sigma [(\overline{\nabla} \cdot (\hat{q}^\dagger \cdot \boldsymbol{u}))_{av}^* (\boldsymbol{n} \cdot \hat{q} \cdot \boldsymbol{v})_{diff}
\end{aligned}
$$

$$+ (\boldsymbol{n} \cdot \hat{q}^\dagger \cdot \boldsymbol{u})^*_{diff} (\overline{\nabla} \cdot (\hat{q} \cdot \boldsymbol{v}))_{av}] \, d\bar{s}$$

$$- \int_{C_2} [(\overline{\nabla} \cdot (\hat{q}^\dagger \cdot \boldsymbol{u}))^*(\boldsymbol{n} \cdot \hat{q} \cdot \boldsymbol{v})$$

$$+ (\boldsymbol{n} \cdot \hat{q}^\dagger \cdot \boldsymbol{u})^*(\overline{\nabla} \cdot (\hat{q} \cdot \boldsymbol{v}))] \, d\bar{s}\} \tag{3.9}$$

where $\boldsymbol{u}, \boldsymbol{v} \in (H^1(\Omega))^3$. Therefore, C^0-elements will suffice.

Now, the two-dimensional domain $\bar{\Omega}$ is meshed into a number of triangular elements. For the work presented in this book, the set of expansion functions in the finite element model are constructed as follows.

For first-order elements, as shown in Figure 3.1(a), the local vector shape functions $\boldsymbol{\psi}^e_r$ are constructed in the form:

$$\boldsymbol{\psi}^e_r \triangleq \begin{cases} \xi_l \boldsymbol{a}_k & k = 1, 2 \\ j\xi_l \boldsymbol{a}_k & k = 3 \end{cases} \qquad r = 3(l-1) + k \text{ (so } r \in \sigma_9), \ \forall l \in \sigma_3 \tag{3.10}$$

whereas for second-order elements, as shown in Figure 3.1(b), the $\boldsymbol{\psi}^e_r$ are defined as

$$\boldsymbol{\psi}^e_r \triangleq \begin{cases} (2\xi_l - 1)\xi_l \boldsymbol{a}_k & k = 1, 2 \\ j(2\xi_l - 1)\xi_l \boldsymbol{a}_k & k = 3, \end{cases} \quad r = 3(l-1) + k \text{ (so } r \in \sigma_9) \\ \qquad\qquad\qquad\qquad\qquad\qquad\qquad\qquad \forall l \in \sigma_3, \\ \begin{cases} 4\xi_l \xi_{l+1} \boldsymbol{a}_k & k = 1, 2 \\ j4\xi_l \xi_{l+1} \boldsymbol{a}_k & k = 3 \end{cases} \quad r = 3(l+2) + k \text{ (so } r \in \sigma_{18} \setminus \sigma_9) \tag{3.11}$$

where $j = \sqrt{-1}$, $\boldsymbol{a}_k$ ($\forall k \in \sigma_3$) are the unit vectors of the global 3D Cartesian coordinates (including $\bar{z}$), and the local area coordinates ξ_l ($\forall l \in \sigma_3$) are defined by [100, 102]

$$\begin{bmatrix} \xi_1 \\ \xi_2 \\ \xi_3 \end{bmatrix} = \frac{1}{2A} \begin{bmatrix} a_1 & b_1 & c_1 \\ a_2 & b_2 & c_2 \\ a_3 & b_3 & c_3 \end{bmatrix} \begin{bmatrix} 1 \\ \bar{x} \\ \bar{y} \end{bmatrix} \tag{3.12}$$

with A being the area of the element given by

$$A = \frac{1}{2} \begin{vmatrix} 1 & 1 & 1 \\ \bar{x}_1 & \bar{x}_2 & \bar{x}_3 \\ \bar{y}_1 & \bar{y}_2 & \bar{y}_3 \end{vmatrix} \tag{3.13}$$

and

$$\begin{aligned} a_l &= \bar{x}_{l+1}\bar{y}_{l-1} - \bar{x}_{l-1}\bar{y}_{l+1} \\ b_l &= \bar{y}_{l+1} - \bar{y}_{l-1} \\ c_l &= \bar{x}_{l-1} - \bar{x}_{l+1} \end{aligned} \tag{3.14}$$

Here, the subscripts $l+1$ and $l-1$ in Equations 3.11 and 3.14 always progress modulo 3, *i.e.*, in cyclic order: 1, 2, 3; and $(\bar{x}_l, \bar{y}_l)$ are the global Cartesian coordinates of vertex l of the triangular element e.

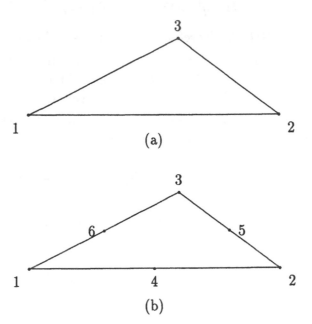

Figure 3.1: Two-dimensional triangular elements: (a) first-order; (b) second-order.

We now introduce the connection matrices Λ^e with elements defined by

$$
\Lambda^e_{i,r} = \begin{cases} 1 & \text{if the global basis function } v_i \text{ is identical to the} \\ & \text{local basis function } \psi^e_r \text{ within element } e \\ 0 & \text{otherwise} \end{cases} \tag{3.15}
$$

so that the global basis function v_i is given by[2]

$$
v_i(\bar{x}, \bar{y}) = \sum_{e=1}^{n_e} v^e_i(\bar{x}, \bar{y}) \qquad (\bar{x}, \bar{y}) \in \Omega \quad \forall i \in \sigma_q \tag{3.16}
$$

where

$$
v^e_i(\bar{x}, \bar{y}) = \begin{cases} \sum_{r=1}^{t} \Lambda^e_{i,r} \, \psi^e_r(\bar{x}, \bar{y}) & \text{for } (\bar{x}, \bar{y}) \in \Omega^e \\ 0 & \text{for } (\bar{x}, \bar{y}) \notin \Omega^e \end{cases} \tag{3.17}
$$

---

[2]The summation in Equation 3.16 is actually a process of patching, *i.e.* the shape functions along interelement boundaries are counted once and only once. A detailed description can be found in [99, pp. 37-40].

Here $q = m \times n_s$ (n_s is the total number of nodes in the model, $m = 3$ is the number of components of the vector field) is the total number of degrees of freedom, and the mapping between the global index of basis functions $i = i(g, k)$ and the local index of basis functions r depends on the details of node numbering for the node-based elements adopted in this book. For the example of first-order elements, if a particular node of element e has local node index $l = l(e, g)$ and global node index $g = g(e, l)$, then

$$v_i^e(\bar{x}, \bar{y}) = \psi_r^e(\bar{x}, \bar{y}) \qquad (\bar{x}, \bar{y}) \in \Omega^e \qquad (3.18)$$

where $i = 3(g - 1) + k$ and $r = 3(l - 1) + k$ ($\forall k \in \sigma_3$).

Up to this point the construction of the finite-dimensional expansion space with v_i ($\forall i \in \sigma_q$) as vector basis functions has been completed. The testing space can be chosen differently from the expansion space even though they belong to the same Hilbert space. Now let the testing and the expansion spaces be the same. Then the general method of moments leads to the Galerkin method, which has the advantage of yielding hermitian or symmetric matrices in the final matrix equation.

It can be shown that the above construction is equivalent to the conventional Galerkin FEM using nodal unknowns, but the technique of constructing vector basis functions for a finite element model adopted in this work is a generalisation, leaving open the development of special vector finite elements for the solution of quasi-3D waveguide problems with the solenoidal constraint for future work.

The form chosen for the expansion and the testing functions in Equations 3.10 and 3.11 is advantageous as the final matrix equation will be real symmetric when the hermitian dyadics $\hat{\epsilon}_r$ and $\hat{\mu}_r$ have the form

$$\hat{\kappa} = \begin{bmatrix} \kappa_{xx} & \kappa_{xy} & \kappa_{xz} \\ \kappa_{yx} & \kappa_{yy} & \kappa_{yz} \\ \kappa_{zx} & \kappa_{zy} & \kappa_{zz} \end{bmatrix} \qquad (3.19)$$

with κ_{xy} being real and κ_{xz} and κ_{yz} being purely imaginary. Consequently, the computations will be much cheaper. Actually, the above form covers most of the lossless optical materials encountered. Of course, one can equivalently make the vector shape functions in Equations 3.10 and 3.11 imaginary in the transverse directions and real in the longitudinal direction.

Now the trial vector field $\widetilde{V}(\bar{x}, \bar{y})$ can be expanded in terms of the vector basis functions $v_j(\bar{x}, \bar{y})$ $(\forall j \in \sigma_q)$:

$$\tilde{V}(\bar{x}, \bar{y}) = \sum_{j=1}^{q} \alpha_j v_j(\bar{x}, \bar{y}) \tag{3.20}$$

where α_j, $\forall j \in \sigma_q$, are the unknown coefficients.

Application of the Galerkin FEM to the operator equation (Equation 2.29) yields

$$< v_i, \ \hat{L}_e \tilde{V} >_e -\lambda < v_i, \ M\tilde{V} >_e$$

$$=< v_i, \ \sum_{j=1}^{q} \alpha_j \hat{L}_e v_j >_e -\lambda < v_i, \ \sum_{j=1}^{q} \alpha_j M v_j >_e$$

$$= \sum_{j=1}^{q} [< v_i, \ \hat{L}_e v_j >_e -\lambda < v_i, \ M v_j >_e] \alpha_j$$

$$= 0 \qquad\qquad\qquad \forall i \in \sigma_q \tag{3.21}$$

Here the subscript e stands for "extended" rather than "element" which is denoted by the superscript e.

If Equation 3.21 is written in a matrix form, the following generalised eigenvalue problem is obtained:

$$[S + pU]\,\alpha - \lambda T\,\alpha = 0 \tag{3.22}$$

with the *normalised propagation constant* $\bar{\beta} = \beta/\rho$ being included in S and U, where α is a $q \times 1$ column vector of unknown coefficients[3] and S, U and T are $q \times q$ matrices[4] which are assembled from the element matrices:

$$[S^e]_{ij} \triangleq \int_{\Omega^e} (\overline{\nabla} \times v_i^e)^* \cdot \hat{p}^{-1} \cdot (\overline{\nabla} \times v_j^e)\, d\bar{x} d\bar{y}$$

$$- \oint_{\Sigma^e} [(\hat{p}^{\dagger-1} \cdot (\overline{\nabla} \times v_i^e))_{av}^* \cdot (n \times v_j^e)_{diff}$$

$$+ (n \times v_i^e)_{diff}^* \cdot (\hat{p}^{-1} \cdot (\overline{\nabla} \times v_j^e))_{av}]\, d\bar{s}$$

$$- \int_{C_1^e} [(\hat{p}^{\dagger-1} \cdot (\overline{\nabla} \times v_i^e))^* \cdot (n \times v_j^e)$$

$$+ (n \times v_i^e)^* \cdot (\hat{p}^{-1} \cdot (\overline{\nabla} \times v_j^e))]\, d\bar{s} \tag{3.23}$$

---

[3]They happen to be nodal values of the components of the trial vector field for vector basis functions constructed from Equation 3.10 and 3.11. In general, they can be abstract expansion coefficients.

[4]These matrices are hermitian for hermitian dyadics $\hat{\epsilon}_r$ and $\hat{\mu}_r$, and become real symmetric when $\hat{\epsilon}_r$ and $\hat{\mu}_r$ satisfy Equation 3.19.

$$[U^e]_{ij} \triangleq \int_{\Omega^e} (\overline{\nabla} \cdot (\hat{q}^\dagger \cdot \boldsymbol{v}_i^e))^* (\overline{\nabla} \cdot (\hat{q} \cdot \boldsymbol{v}_j^e)) \, d\bar{x} d\bar{y}$$

$$- \oint_{\Sigma^e} [(\overline{\nabla} \cdot (\hat{q}^\dagger \cdot \boldsymbol{v}_i^e))^*_{av} (\boldsymbol{n} \cdot \hat{q} \cdot \boldsymbol{v}_j^e)_{diff}$$

$$+ (\boldsymbol{n} \cdot \hat{q}^\dagger \cdot \boldsymbol{v}_i^e)^*_{diff} (\overline{\nabla} \cdot (\hat{q} \cdot \boldsymbol{v}_j^e))_{av}] \, d\bar{s}$$

$$- \int_{C_2^e} [(\overline{\nabla} \cdot (\hat{q}^\dagger \cdot \boldsymbol{v}_i^e))^* (\boldsymbol{n} \cdot \hat{q} \cdot \boldsymbol{v}_j^e)$$

$$+ (\boldsymbol{n} \cdot \hat{q}^\dagger \cdot \boldsymbol{v}_i^e)^* (\overline{\nabla} \cdot (\hat{q} \cdot \boldsymbol{v}_j^e))] \, d\bar{s} \qquad (3.24)$$

and

$$[T^e]_{ij} \triangleq \int_{\Omega^e} (\boldsymbol{v}_i^e)^* \cdot \hat{q} \cdot \boldsymbol{v}_j^e \, d\bar{x} d\bar{y} \qquad (3.25)$$

Here the superscript e again stands for "element", $\Sigma^e = \Omega^e \cap \Sigma$, $C_i^e = \Omega^e \cap C_i$ ($i = 1, 2$), and it is assumed that the two-dimensional domain $\bar{\Omega}$ is meshed in such a way that the interface Σ and the boundary $C = C_1 \cup C_2$ belong to the collection of element boundaries and interelement interfaces in $\bar{\Omega}$. It should be mentioned that Equations 3.23 to 3.25 are only for the convenience of description. In practice, the element matrices are formed in terms of local indices and the relevant unknowns of that element, followed by embedding of those element matrices into the global matrices in Equation 3.22, which is a standard procedure [91, 102] and is omitted for brevity. Moreover, $\hat{p}$ and $\hat{q}$ are problem-dependent, and in FEM programming $\hat{p}^{-1}$ and $\hat{q}$ are interpolated in terms of some parameters, *e.g.* nodal values, on an element-by-element basis so that the package can be applied to problems having different $\hat{p}$ and $\hat{q}$ without the user's intervention to the source code. This arrangement renders numerical integration unnecessary in the computation of the element matrices. Of course, the user has to prepare data files of those parameters for interpolation, which is quite an easy and small task.

Again, the line integral terms in Equations 3.23 and 3.24 will vanish whenever all the discontinuity and boundary conditions can be enforced explicitly.

3.4 Enforcing the Interface and Boundary Conditions

As discussed in § 2.5, it is preferable for expansion and testing functions to satisfy *a priori* as many of the required interface and boundary conditions as possible for best numerical results even though that is unnecessary in the extended operator formulation.

In the intermediate form of the extended operator (Equation 3.9), only some of the discontinuity and boundary conditions specified in Equation 2.2 and 2.33 appear explicitly. These are called *essential* discontinuity and boundary conditions; the others are called *natural*. There is no need for the expansion and testing functions to satisfy the natural conditions nor the essential conditions when using the extended operator, but the corresponding interface and boundary integrals will vanish if each vector basis function, *e.g.* v_i ($i \in \sigma_q$), exactly satisfies the essential discontinuity and boundary conditions, namely:

$$\begin{cases} \boldsymbol{n} \times \boldsymbol{v}_i = \boldsymbol{0} & \text{on } C_1 \\ \boldsymbol{n} \cdot \hat{q} \cdot \boldsymbol{v}_i = 0 & \text{on } C_2 \\ \\ (\boldsymbol{n} \times \boldsymbol{v}_i)_{diff} = \boldsymbol{0} & \text{on } \Sigma \\ (\boldsymbol{n} \cdot \hat{q} \cdot \boldsymbol{v}_i)_{diff} = 0 & \text{on } \Sigma \end{cases} \tag{3.26}$$

This not only makes the computation of the element matrices in Equations 3.23 and 3.24 much simpler, but also improves the numerical solutions.

3.4.1   Interface conditions

In the conventional vector FEM, every component of v_i ($\forall i \in \sigma_q$) is continuous across interelement interfaces and the second interface condition in Equation 3.26 will be violated if $\hat{q}$ is discontinuous across Σ. The known methods of dealing with the problem, suggested in [13] and [14], are only applicable to isotropic cases. It is interesting that there has not been any reported method, at least known to the author, to enforce the discontinuity condition with arbitrary anisotropic $\hat{q}$ except the extended operator method proposed in [45, 65] though many papers have been published on the $\boldsymbol{E}$-vector formulation with discontinuous $\hat{\epsilon}_r$, *e.g.* [13]-[15], [47, 103, 104]. Most probably, the discontinuity problem has been ignored by most authors, and so the conclusion has been reached that the $\boldsymbol{E}$-vector formulation is less accurate than the $\boldsymbol{H}$-vector one [89, 105]. Instead of enforcing interface conditions in the average sense by the extended operator method, an explicit method, suitable for anisotropic media, is proposed in the following.

With the vector basis functions constructed in the previous section, the tangential continuity of v_i ($\forall i \in \sigma_q$) across Σ is already satisfied, and only the continuity of the normal component of $\hat{q} \cdot v_i$ ($\forall i \in \sigma_q$) across Σ needs to be enforced, which can be done easily in isotropic media with *regular interfaces*[5]. In arbitrary anisotropic media or isotropic media with irregular interfaces, however, it may be more convenient to consider both interface conditions as one entity.

For simplicity, consider two adjacent first-order finite elements, Ω^a and Ω^b, with common straight interface $\Sigma^{ab} \triangleq \Omega^a \cap \Omega^b$ on Σ, as illustrated in Figure 3.2. The idea is to begin with the vector shape functions on each side of Σ^{ab} and adjust them so that the interface conditions are satisfied at each of the two nodes on Σ^{ab}. For first-order vector shape functions, a tangential component of v_i ($i \in \sigma_q$) or a normal component of $\hat{q} \cdot v_i$ ($i \in \sigma_q$) will be continuous[6] on the whole of Σ^{ab} if it is continuous at each node on Σ^{ab}. Consequently, that component will be continuous on the whole of Σ if it is continuous at every node on Σ. Therefore, the discussion can now be focussed on how to enforce the required interface conditions at a single node with global index h and global coordinates $(\bar{x}_h, \bar{y}_h)$ on Σ.

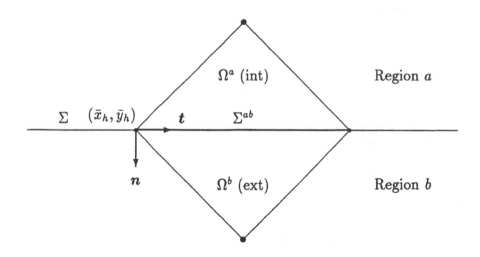

Figure 3.2: Two adjacent first-order finite elements Ω^a and Ω^b with common interface Σ^{ab} on Σ.

---

[5]These are interfaces whose segments are perpendicular to either the x- or the y-axis.

[6]Subject to some approximation when $\hat{q}^a$ and $\hat{q}^b$ are not constant dyadics along Σ^{ab}.

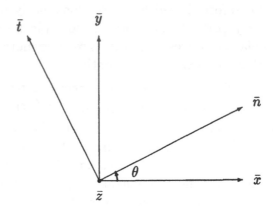

Figure 3.3: The relation between Cartesian global and local coordinate systems.

It is convenient to consider a local coordinate system $(\bar{n}, \bar{t}, \bar{z})$ at the node $(\bar{x}_h, \bar{y}_h)$ on Σ^{ab} as shown in Figure 3.3, where $\theta \in [0, 2\pi)$ is the angle of the unit vector $\boldsymbol{a}_n$ normal[7] to Σ^{ab} with respect to the unit vector $\boldsymbol{a}_x$ in the positive $\bar{x}$ direction. Hence

$$\boldsymbol{n} \stackrel{\triangle}{=} \boldsymbol{a}_n = n_x \boldsymbol{a}_x + n_y \boldsymbol{a}_y = \cos\theta \boldsymbol{a}_x + \sin\theta \boldsymbol{a}_y \qquad (3.27)$$

and

$$\boldsymbol{n} \cdot \boldsymbol{n} = n_x n_x + n_y n_y = 1. \qquad (3.28)$$

The essential discontinuity conditions on Σ given in Equation 3.26 can be expressed as a single dyadic equation

$$(\hat{Q} \cdot \boldsymbol{v}_i)_{diff} = \hat{Q}^a \cdot \boldsymbol{v}_i|^a - \hat{Q}^b \cdot \boldsymbol{v}_i|^b = 0 \qquad \text{on } \Sigma \qquad (3.29)$$

where the dyadic function $\hat{Q}$ is defined as

$$\begin{aligned} \hat{Q} &\stackrel{\triangle}{=} \hat{I} - \boldsymbol{nn} + \boldsymbol{nn} \cdot \hat{q} \\ &= -\boldsymbol{n} \times (\boldsymbol{n} \times \hat{I}) + \boldsymbol{nn} \cdot \hat{q} \end{aligned} \qquad (3.30)$$

and $\hat{I} = \boldsymbol{a}_x \boldsymbol{a}_x + \boldsymbol{a}_y \boldsymbol{a}_y + \boldsymbol{a}_z \boldsymbol{a}_z$ is the identity dyadic. For convenience the global components of the vector function $\boldsymbol{n} \cdot \hat{q}$ will be written as

$$(\boldsymbol{n} \cdot \hat{q})_x \equiv q_{nx} = n_x q_{xx} + n_y q_{yx}$$

---

[7]This is the same as the outward unit normal vector $\boldsymbol{n}$ defined in Equation 2.9.

$$(\boldsymbol{n} \cdot \hat{q})_y \equiv q_{ny} = n_x q_{xy} + n_y q_{yy} \tag{3.31}$$
$$(\boldsymbol{n} \cdot \hat{q})_z \equiv q_{nz} = n_x q_{xz} + n_y q_{yz}$$

Hence the global components of the dyadic $\hat{Q}$ can be displayed as the matrix

$$[Q] = \begin{bmatrix} n_x q_{nx} + n_y n_y & n_x q_{ny} - n_x n_y & n_x q_{nz} \\ n_y q_{nx} - n_y n_x & n_y q_{ny} + n_x n_x & n_y q_{nz} \\ 0 & 0 & 1 \end{bmatrix}. \tag{3.32}$$

Superscripts are attached to the dyadics $\hat{q}$ and $\hat{Q}$ to denote one side of the interface Σ or the other.

The inverse of the dyadic $\hat{Q}$, written $\hat{Q}^{-1}$, is defined such that

$$\hat{Q}^{-1} \cdot \hat{Q} = \hat{Q} \cdot \hat{Q}^{-1} = \hat{I}. \tag{3.33}$$

It is easily shown that

$$\hat{Q}^{-1} = \hat{I} + \frac{\boldsymbol{nn} - \boldsymbol{nn} \cdot \hat{q}}{\boldsymbol{n} \cdot \hat{q} \cdot \boldsymbol{n}} \tag{3.34}$$

and

$$(\hat{Q}^b)^{-1} \cdot (\hat{Q}^a) = \hat{I} + \frac{\boldsymbol{nn} \cdot (\hat{q}^a - \hat{q}^b)}{\boldsymbol{n} \cdot \hat{q}^b \cdot \boldsymbol{n}}. \tag{3.35}$$

Since first-order vector shape functions are being considered, the interface conditions in Equation 3.29 can be approximated by

$$\hat{Q}^a(\bar{x}_h, \bar{y}_h) \cdot \boldsymbol{v}_j(\bar{x}, \bar{y})|^a = \hat{Q}^b(\bar{x}_h, \bar{y}_h) \cdot \boldsymbol{v}_j(\bar{x}, \bar{y})|^b$$
$$\text{for } (\bar{x}, \bar{y}) \text{ on } \Sigma^{ab} \tag{3.36}$$

provided that the dyadic functions $\hat{Q}^a(\bar{x}, \bar{y})$ and $\hat{Q}^b(\bar{x}, \bar{y})$ do not vary significantly along Σ^{ab}.

A general method of constructing a modified vector basis function $\boldsymbol{v}_i(\bar{x}, \bar{y})$ satisfying the interface conditions is to patch together the corresponding vector shape functions which have been individually adjusted by an appropriate dyadic transformation to account for Equation 3.36, that is,

$$\boldsymbol{v}_i(\bar{x}, \bar{y}) = \begin{cases} \hat{A} \cdot \boldsymbol{v}_i^a(\bar{x}, \bar{y}) & \text{for } (\bar{x}, \bar{y}) \in \Omega^a \\ \hat{B} \cdot \boldsymbol{v}_i^b(\bar{x}, \bar{y}) & \text{for } (\bar{x}, \bar{y}) \in \Omega^b \\ \vdots & \vdots \end{cases} \tag{3.37}$$

where

$$v_i^a(\bar{x},\bar{y}) \;=\; v_i^b(\bar{x},\bar{y}) \qquad \text{for } (\bar{x},\bar{y}) \text{ on } \Sigma^{ab} \tag{3.38}$$

and

$$\hat{Q}^a(\bar{x}_h,\bar{y}_h) \cdot \hat{A} \;=\; \hat{Q}^b(\bar{x}_h,\bar{y}_h) \cdot \hat{B}. \tag{3.39}$$

Here $v_i^a(\bar{x},\bar{y})$ and $v_i^b(\bar{x},\bar{y})$ are the vector shape functions $v_i^e(\bar{x},\bar{y})$ constructed in § 3.3.

The constant dyadics, $\hat{A}$ and $\hat{B}$, are not uniquely defined by Equation 3.39, which represents the essential discontinuity conditions, until another relation is imposed which determines the amplitude of the vector basis function $v_i(\bar{x},\bar{y})$ and hence the scale of the nodal coefficient α_i.

Now the approximate electromagnetic field (after solving by the FEM) is

$$
\begin{aligned}
\tilde{V}(\bar{x},\bar{y}) \;&=\; \sum_{i=1}^{q} \alpha_i v_i(\bar{x},\bar{y}) \\[2mm]
&=\;
\begin{cases}
\sum_{i \,(\text{on } \Sigma^{ab})} \alpha_i \hat{A} \cdot v_i^a(\bar{x},\bar{y}) & \text{for } (\bar{x},\bar{y}) \in \Omega^a \\[2mm]
\sum_{i \,(\text{on } \Sigma^{ab})} \alpha_i \hat{B} \cdot v_i^b(\bar{x},\bar{y}) & \text{for } (\bar{x},\bar{y}) \in \Omega^b \\[1mm]
\;\vdots & \qquad \vdots
\end{cases}
\end{aligned}
\tag{3.40}
$$

for points $(\bar{x},\bar{y})$ just on either side of the interface Σ^{ab}. In particular, at the node $(\bar{x}_h,\bar{y}_h)$ on Σ^{ab}, the two values of the vector field on each side of Σ^{ab} are

$$\tilde{V}^a(\bar{x}_h,\bar{y}_h) \;=\; \hat{A} \cdot \left(\sum_{i \,(\text{at node } h)} \alpha_i v_i^a(\bar{x}_h,\bar{y}_h) \right) \tag{3.41}$$

$$\tilde{V}^b(\bar{x}_h,\bar{y}_h) \;=\; \hat{B} \cdot \left(\sum_{i \,(\text{at node } h)} \alpha_i v_i^b(\bar{x}_h,\bar{y}_h) \right) \tag{3.42}$$

Since the vector shape functions, $v_i^a(\bar{x},\bar{y})$ and $v_i^b(\bar{x},\bar{y})$, have been constructed according to Equations 3.10 and 3.17, they satisfy

$$v_i^e(\bar{x}_h,\bar{y}_h) = \begin{cases} \eta_k\,a_k & \text{when } i = 3(h-1)+k; \; k = 1,2,3 \\ 0\,a_k & \text{otherwise} \end{cases} \tag{3.43}$$

where

$$\eta_k = \begin{cases} 1 & k = 1, 2 \\ j & k = 3 \end{cases} \tag{3.44}$$

with $j = \sqrt{-1}$. Hence

$$\widetilde{V}^a(\bar{x}_h, \bar{y}_h) = \hat{A} \cdot \left(\sum_{k=1}^{3} \alpha_{3(h-1)+k} \, \eta_k \boldsymbol{a}_k \right) \tag{3.45}$$

$$\widetilde{V}^b(\bar{x}_h, \bar{y}_h) = \hat{B} \cdot \left(\sum_{k=1}^{3} \alpha_{3(h-1)+k} \, \eta_k \boldsymbol{a}_k \right) \tag{3.46}$$

where the sum in the brackets represents the "nodal vector-field value". These two equations indicate how appropriate choices of the constant dyadics, $\hat{A}$ and $\hat{B}$, might be made. Three useful examples follow, where the first example is the technique used for the subsequent analysis and in the computer program MEF.FOR in Appendix B.3.

Example 1

$$\hat{A} = \hat{I}, \qquad \hat{B} = (\hat{Q}^b)^{-1} \cdot (\hat{Q}^a). \tag{3.47}$$

Then

$$\sum_{k=1}^{3} \alpha_{3(h-1)+k} \, \eta_k \boldsymbol{a}_k = \widetilde{V}^a(\bar{x}_h, \bar{y}_h)$$

$$= (\hat{Q}^a)^{-1} \cdot (\hat{Q}^b) \cdot \widetilde{V}^b(\bar{x}_h, \bar{y}_h). \tag{3.48}$$

Here the nodal coefficients, $\alpha_{3(h-1)+k} \, \eta_k$ $(k = 1, 2, 3)$, represent the nodal values of the three components of the vector field on the Ω^a side of the interface at the node h.

Alternatively, one can choose

$$\hat{A} = (\hat{Q}^a)^{-1} \cdot (\hat{Q}^b), \qquad \hat{B} = \hat{I}. \tag{3.49}$$

Consequently,

$$\sum_{k=1}^{3} \alpha_{3(h-1)+k} \, \eta_k \boldsymbol{a}_k = (\hat{Q}^b)^{-1} \cdot (\hat{Q}^a) \cdot \widetilde{V}^a(\bar{x}_h, \bar{y}_h)$$

$$= \widetilde{V}^b(\bar{x}_h, \bar{y}_h). \tag{3.50}$$

Now the nodal coefficients, $\alpha_{3(h-1)+k}\,\eta_k$ $(k = 1, 2, 3)$, represent the nodal values of the three components of the vector field on the Ω^b side of the interface at the node h.

Example 2

$$\hat{A} = (\hat{Q}^a)^{-1}, \qquad \hat{B} = (\hat{Q}^b)^{-1}. \tag{3.51}$$

Then

$$\sum_{k=1}^{3} \alpha_{3(h-1)+k}\,\eta_k a_k \;=\; \hat{Q}^a \cdot \tilde{V}^a(\bar{x}_h, \bar{y}_h)$$

$$\;=\; \hat{Q}^b \cdot \tilde{V}^b(\bar{x}_h, \bar{y}_h). \tag{3.52}$$

Note that this choice has the advantage that only the local value of $\hat{q}$ is used to construct the modified vector basis functions. However, the normal component of the nodal vector-field value is the normal component of the "relative flux density" (D/ϵ_0 or B/μ_0) rather than the "field intensity" (E or H) itself.

Example 3

$$\hat{A} \;=\; 2(\hat{Q}^a + \hat{Q}^b)^{-1} \cdot \hat{Q}^b = 2[\hat{I} + (\hat{Q}^b)^{-1} \cdot \hat{Q}^a]^{-1},$$
$$\hat{B} \;=\; 2(\hat{Q}^a + \hat{Q}^b)^{-1} \cdot \hat{Q}^a = 2[\hat{I} + (\hat{Q}^a)^{-1} \cdot \hat{Q}^b]^{-1}. \tag{3.53}$$

Then

$$\sum_{k=1}^{3} \alpha_{3(h-1)+k}\,\eta_k a_k \;=\; \tfrac{1}{2}[\hat{I} + (\hat{Q}^b)^{-1} \cdot \hat{Q}^a] \cdot \tilde{V}^a(\bar{x}_h, \bar{y}_h)$$

$$\;=\; \tfrac{1}{2}[\hat{I} + (\hat{Q}^a)^{-1} \cdot \hat{Q}^b] \cdot \tilde{V}^b(\bar{x}_h, \bar{y}_h)$$
$$\;=\; \tfrac{1}{2}[\tilde{V}^a(\bar{x}_h, \bar{y}_h) + \tilde{V}^b(\bar{x}_h, \bar{y}_h)]$$
$$\;=\; \tilde{V}_{av}(\bar{x}_h, \bar{y}_h) \tag{3.54}$$

If discontinuous fields cannot be plotted by the available resources, this choice has the advantage that the nodal vector-field value is the average value of the field at an interface.

For reference, the dyadics appearing in Example 3 are

$$\hat{A} \;=\; 2(\hat{Q}^a + \hat{Q}^b)^{-1} \cdot \hat{Q}^b = 2[\hat{I} + (\hat{Q}^b)^{-1} \cdot \hat{Q}^a]^{-1}$$
$$\;=\; \hat{I} + \frac{nn \cdot (\hat{q}^b - \hat{q}^a)}{n \cdot (\hat{q}^a + \hat{q}^b) \cdot n} = \left[\hat{I} + \frac{nn \cdot (\hat{q}^a - \hat{q}^b)}{2n \cdot \hat{q}^b \cdot n}\right]^{-1}$$

$$(3.55)$$

$$\hat{B} = 2(\hat{Q}^a + \hat{Q}^b)^{-1} \cdot \hat{Q}^a = 2[\hat{I} + (\hat{Q}^a)^{-1} \cdot \hat{Q}^b]^{-1}$$

$$= \hat{I} + \frac{\boldsymbol{nn} \cdot (\hat{q}^a - \hat{q}^b)}{\boldsymbol{n} \cdot (\hat{q}^a + \hat{q}^b) \cdot \boldsymbol{n}} = \left[\hat{I} + \frac{\boldsymbol{nn} \cdot (\hat{q}^b - \hat{q}^a)}{2\boldsymbol{n} \cdot \hat{q}^a \cdot \boldsymbol{n}}\right]^{-1} \quad (3.56)$$

Moreover, it is noted that in this example

$$\hat{A} + \hat{B} \equiv 2\hat{I}. \quad (3.57)$$

The use of such modified vector basis functions for both expansion and testing functions does not disturb the hermitian or real symmetric property of the resulting algebraic matrix equation.

It can be shown that the above technique of explicitly enforcing the interface conditions is exact when $\hat{q}^a$ and $\hat{q}^b$ are constant dyadics on Σ^{ab}, and is only an approximation when $\hat{q}^a$ and $\hat{q}^b$ are inhomogeneous on Σ^{ab}. The concepts and methodology presented here can be extended to higher-order vector shape functions, the above discussion having been confined to first-order vector shape functions for simplicity.

3.4.2 Boundary conditions

It is preferable that the essential boundary conditions are enforced explicitly as well, though it is not mandatory when employing the extended operator. As shown in Equation 3.26, the essential boundary conditions are of homogeneous Dirichlet type and can be expressed as a pair of dyadic equations

$$\begin{aligned}
\boldsymbol{n} \times (\hat{Q} \cdot \boldsymbol{v}_i) &= \boldsymbol{0} &\quad \text{on } C_1 \\
\boldsymbol{n} \cdot (\hat{Q} \cdot \boldsymbol{v}_i) &= 0 &\quad \text{on } C_2
\end{aligned} \quad (3.58)$$

or

$$\begin{aligned}
(\hat{I} - \boldsymbol{nn}) \cdot (\hat{Q} \cdot \boldsymbol{v}_i) &= \boldsymbol{0} &\quad \text{on } C_1 \\
\boldsymbol{nn} \cdot (\hat{Q} \cdot \boldsymbol{v}_i) &= \boldsymbol{0} &\quad \text{on } C_2
\end{aligned} \quad (3.59)$$

where the dyadic function $\hat{Q}$ has been defined in Equation 3.30, and is evaluated at the appropriate nodal points.

As in the previous subsection, a modified vector basis function $\boldsymbol{v}_i$ satisfying the boundary conditions on either C_1 or C_2 can be constructed by patching together the corresponding vector shape functions which have been individually adjusted by an appropriate dyadic transformation to account for Equation 3.59:

$$v_i(\bar{x}, \bar{y}) = \begin{cases} \hat{C} \cdot v_i^c(\bar{x}, \bar{y}) & \text{for } (\bar{x}, \bar{y}) \in \Omega^c \\ \vdots & \vdots \end{cases} \qquad (3.60)$$

where either

$$(\hat{I} - nn) \cdot (\hat{Q}^c(\bar{x}_h, \bar{y}_h) \cdot \hat{C}) = (\hat{I} - nn) \cdot \hat{C} = \hat{0} \qquad \text{on } C_1 \qquad (3.61)$$

or

$$nn \cdot (\hat{Q}^c(\bar{x}_h, \bar{y}_h) \cdot \hat{C}) = nn \cdot \hat{q}^c(\bar{x}_h, \bar{y}_h) \cdot \hat{C} = \hat{0} \qquad \text{on } C_2 \qquad (3.62)$$

Here Ω^c denotes the domain of a first-order element which has one edge on the boundary[8], one vertex of which is the boundary node $(\bar{x}_h, \bar{y}_h)$.

The constant dyadic $\hat{C}$ is not uniquely defined by either Equation 3.61 or 3.62 until another relation is imposed which determines the amplitude of the vector basis function $v_i(\bar{x}, \bar{y})$ and hence the scale of the nodal coefficient α_i. At the node $(\bar{x}_h, \bar{y}_h)$ on either C_1 or C_2, the approximate electromagnetic field (after solving by the FEM) is

$$\tilde{V}^c(\bar{x}_h, \bar{y}_h) = \hat{C} \cdot \left(\sum_{k=1}^{3} \alpha_{3(h-1)+k} \, \eta_k a_k \right) \qquad (3.63)$$

where the sum in brackets represents the "nodal vector-field value", and η_k has been defined in Equation 3.44. Useful choices of $\hat{C}$ are

$$\hat{C} = \hat{C}_1 = (\hat{Q}^c)^{-1} \cdot nn = \frac{nn}{n \cdot \hat{q}^c \cdot n} \qquad \text{on } C_1 \qquad (3.64)$$

$$\hat{C} = \hat{C}_2 = (\hat{Q}^c)^{-1} \cdot (\hat{I} - nn) = \hat{I} - \frac{nn \cdot \hat{q}^c}{n \cdot \hat{q}^c \cdot n} \qquad \text{on } C_2 \qquad (3.65)$$

Because these dyadics are not invertible, there are no longer three degrees of freedom for the nodal coefficients in Equation 3.63. This can be seen by expanding Equation 3.63 into

$$\tilde{V}^c(\bar{x}_h, \bar{y}_h) = \sum_{k=1}^{3} \alpha_{3(h-1)+k} \, \eta_k n \frac{n \cdot a_k}{n \cdot \hat{q}^c \cdot n}$$

---

[8]A first-order element which has two edges on the boundary is called a *trivial element* in this work. Trivial elements are totally banned from formation in Chapter 4.

$$= \frac{1}{\boldsymbol{n} \cdot \hat{q}^c \cdot \boldsymbol{n}} \left\{ \alpha_{3(h-1)+1} \, n_x + \alpha_{3(h-1)+2} \, n_y \right\} \boldsymbol{n} \quad \text{on } C_1 \quad (3.66)$$

$$\tilde{\boldsymbol{V}}^c(\bar{x}_h, \bar{y}_h) = \sum_{k=1}^{3} \alpha_{3(h-1)+k} \, \eta_k \left(\boldsymbol{a}_k - \boldsymbol{n} \frac{\boldsymbol{n} \cdot \hat{q}^c \cdot \boldsymbol{a}_k}{\boldsymbol{n} \cdot \hat{q}^c \cdot \boldsymbol{n}} \right)$$

$$= \sum_{k=1}^{3} \alpha_{3(h-1)+k} \, \eta_k \frac{\boldsymbol{n} \cdot \hat{q}^c \times (\boldsymbol{a}_k \times \boldsymbol{n})}{\boldsymbol{n} \cdot \hat{q}^c \cdot \boldsymbol{n}}$$

$$= \frac{1}{\boldsymbol{n} \cdot \hat{q}^c \cdot \boldsymbol{n}} \left\{ (\alpha_{3(h-1)+1} \, n_y - \alpha_{3(h-1)+2} \, n_x)(q_{ny}^c \boldsymbol{a}_x - q_{nx}^c \boldsymbol{a}_y) \right.$$
$$\left. + j\alpha_{3(h-1)+3} \left(\boldsymbol{n} \cdot \hat{q}^c \cdot \boldsymbol{n} \, \boldsymbol{a}_z - q_{nz}^c \boldsymbol{n} \right) \right\} \qquad \text{on } C_2 \quad (3.67)$$

The vector field in Equation 3.66 is directed normal to the boundary C_1 and has only one degree of freedom, while the vector field in Equation 3.67 spans the two-dimensional space that is perpendicular to the direction of $\boldsymbol{n} \cdot \hat{q}^c$ on the boundary C_2 and has two degrees of freedom.

If the boundary segments are perpendicular to either the x-axis or the y-axis, so that $\boldsymbol{n} = \pm \boldsymbol{a}_x$ or $\pm \boldsymbol{a}_y$, then the excess degrees of freedom are easy to remove. Suppose $\boldsymbol{n} = \boldsymbol{a}_x$, for example. Then Equations 3.66 and 3.67 become

$$\tilde{\boldsymbol{V}}^c(\bar{x}_h, \bar{y}_h) = \frac{1}{q_{xx}^c} \alpha_{3(h-1)+1} \, \boldsymbol{a}_x \qquad \text{on } C_1 \qquad (3.68)$$

$$\tilde{\boldsymbol{V}}^c(\bar{x}_h, \bar{y}_h) = \frac{1}{q_{xx}^c} \left\{ -\alpha_{3(h-1)+2} \left(q_{xy}^c \boldsymbol{a}_x - q_{xx}^c \boldsymbol{a}_y \right) \right.$$
$$\left. + j\alpha_{3(h-1)+3} \left(q_{xx}^c \boldsymbol{a}_z - q_{xz}^c \boldsymbol{a}_x \right) \right\} \qquad \text{on } C_2 \quad (3.69)$$

The missing nodal coefficients α_i are set to zero. In order not to rearrange the global matrix equation (Equation 3.22), the ith equation should be replaced by

$$\gamma_i \alpha_i = \lambda \zeta_i \alpha_i \qquad i \in \sigma_q \qquad (3.70)$$

and also the coefficients in the ith column of each matrix of the global matrix equation should be set to zero except the diagonal one [101, 102, 106]. Here γ_i and ζ_i are some appropriate constants dependent on the range of the eigenvalues to be found and the associated solution method. For example, if the solution algorithm adopted is to extract the smallest eigenvalue λ_{min} and sup $\lambda_{min} < 10$, then set $\gamma_i/\zeta_i > 10$ so that the smallest eigenvalue corresponds to a physical solution rather than to γ_i/ζ_i.

For boundary segments whose normals are at arbitrary angles with respect to the global Cartesian axes, the excess degrees of freedom in

Equations 3.66 and 3.67 are eliminated by introducing homogeneous linear constraints between the nodal coefficients:

$$n_y \alpha_{3(h-1)+1} - n_x \alpha_{3(h-1)+2} = 0 \qquad \text{on } C_1 \qquad (3.71)$$
$$\alpha_{3(h-1)+3} = 0$$

$$n_x \alpha_{3(h-1)+1} + n_y \alpha_{3(h-1)+2} = 0 \qquad \text{on } C_2 \qquad (3.72)$$

Equations 3.71 and 3.72 are *coupled* Dirichlet boundary conditions, and can be used to eliminate the unwanted nodal coefficients explicitly. It is noted, however, that the insertion of coupled Dirichlet boundary conditions by the elimination technique causes the matrix in the global matrix equation (Equation 3.22 in our case) to be no longer real symmetric or hermitian, which is a disadvantage for eigenvalue problems. This can be circumvented by constructing vector basis functions using the local Cartesian coordinate system $(\bar{n}, \bar{t}, \bar{z})$ such that the relevant nodal coefficients represent the tangential components of the vector field $\widetilde{V}$ or the normal component of $\hat{q} \cdot \widetilde{V}$. Consequently, the coupled Dirichlet boundary conditions will be uncoupled.

It should be mentioned that when a nodal coefficient α_i is eliminated or set to zero, the corresponding vector shape function v_i is not required for expansion nor for testing.

Finally, after the constraints in Equations 3.71 and 3.72 have been imposed, the approximate electromagnetic field at a boundary node h has one of the following forms:

$$\widetilde{V}^c(\bar{x}_h, \bar{y}_h) = \frac{1}{n \cdot \hat{q}^c \cdot n} \frac{\alpha_{3(h-1)+1}}{n_x} n \qquad \text{on } C_1 \qquad (3.73)$$

with $\alpha_{3(h-1)+2} = \frac{n_y}{n_x} \alpha_{3(h-1)+1}$ and $\alpha_{3(h-1)+3} = 0$.

$$\widetilde{V}^c(\bar{x}_h, \bar{y}_h) = \frac{1}{n \cdot \hat{q}^c \cdot n} \frac{\alpha_{3(h-1)+2}}{n_y} n \qquad \text{on } C_1 \qquad (3.74)$$

with $\alpha_{3(h-1)+1} = \frac{n_x}{n_y} \alpha_{3(h-1)+2}$ and $\alpha_{3(h-1)+3} = 0$.

$$\widetilde{V}^c(\bar{x}_h, \bar{y}_h) = \frac{1}{n \cdot \hat{q}^c \cdot n} \Big\{ \frac{\alpha_{3(h-1)+1}}{n_y} (q^c_{ny} a_x - q^c_{nx} a_y)$$
$$+ j\alpha_{3(h-1)+3} (n \cdot \hat{q}^c \cdot n\, a_z - q^c_{nz} n) \Big\} \quad \text{on } C_2 \quad (3.75)$$

with $\alpha_{3(h-1)+2} = -\frac{n_x}{n_y} \alpha_{3(h-1)+1}$.

$$\widetilde{V}^c(\bar{x}_h, \bar{y}_h) = \frac{1}{\boldsymbol{n} \cdot \hat{q}^c \cdot \boldsymbol{n}} \{ -\frac{\alpha_{3(h-1)+2}}{n_x} (q_{ny}^c \boldsymbol{a}_x - q_{nx}^c \boldsymbol{a}_y) $$
$$+ j\alpha_{3(h-1)+3} \left(\boldsymbol{n} \cdot \hat{q}^c \cdot \boldsymbol{n}\, \boldsymbol{a}_z - q_{nz}^c \boldsymbol{n}\right)\} \quad \text{on } C_2 \quad (3.76)$$

with $\alpha_{3(h-1)+1} = -\frac{n_y}{n_x} \alpha_{3(h-1)+2}$.

As in the previous subsection, the technique of explicitly enforcing the boundary conditions presented here is exact when $\hat{q}^c$ is a constant dyadic on C_1 or C_2, and is only an approximation when $\hat{q}^c$ is inhomogeneous on the relevant boundary. This technique can also be extended to higher-order vector shape functions.

3.4.3 Singularity at corners

A singularity would arise at any corners of the boundaries or the interfaces between different media. At such singular points the interface and boundary conditions cannot be satisfied exactly as the normal component and one of the tangential components are indefinite. Improper modelling of a singularity may result in poor convergence of results. It may be possible to incorporate some analytical approach or singularity elements into our FEM model to simulate a singularity locally at such points on a case-by-case basis, as is done in [60, 102, 107, 108]. However, such a problem-dependent approach prevents it from being applied to any robust[9] and user-friendly package as the one being developed here.

Being unable to define the normal component and the tangential component in the transverse direction at singular points, we simply make every component of $\boldsymbol{v}_i$ ($\forall i \in \sigma_q$) continuous there in order to achieve robustness at the expense of accuracy. Fortunately, the expense is usually negligible globally as compared with the error resulting from the relatively large size of elements used in the computations, but it may have some local effects.

---

[9]For the definition of robustness, we refer to [109].

Chapter 4

Automatic Mesh Generation

4.1 Introduction

As long as the FEM is the main technique for numerical approximate solution of partial differential-equations which govern the continua, automatic mesh generation has obviously a practical value in reducing errors and the time involved in data preparation. A number of 2D mesh generation algorithms of varying degrees of automation have been proposed [110]. Since triangular elements have certain advantages over isoparametric ones [111, 112], quite a number of algorithms have been suggested for generating a triangular mesh within the given planar domain [113]-[118]. Of all these techniques, the only general approach is automatic triangulation which possesses several features that usually cannot be found in other strategies [112]. Thus, triangular elements are chosen as the building block for the finite element approach in the present work. Having investigated existing techniques, the author developed a new version of automatic mesh generation scheme for producing first- and second-order triangular elements, which is believed to be useful, especially for waveguide structures.

A good scheme for node generation makes the mesh generation straightforward. Such a scheme should yield uniformly distributed nodes directly instead of going through a generating-testing-removing procedure.

There is no difficulty in writing a mesh generation program which is capable of meshing a closed planar domain with any topology as long as one can define it. But such a program can be lengthy, inefficient,

51

fragile and "unfriendly" for users besides involving a lot of tedious programming. Fortunately, the interfaces of optical waveguide structures are often regular, and open boundaries are generally the case. Since only guided waves are of interest, the closed boundary approximation can be adopted and would be sufficient for most applications. The artificial closed boundary can be a simple regular shape. Thus, the structures considered here are restricted to rectangles and circles, corresponding to integrated optics and fibre structures, respectively, or various combinations of these; but the algorithm itself in principle can easily be extended to cases with irregular interfaces and boundaries. As for structures other than the above, they will be treated as entirely inhomogeneous problems in the FEM system rather than being homogeneous region by region, *i.e.* the mesh is generated in the usual way, making the interfaces compatible with the mesh where possible, but the "load" (permittivity) at every node of each element is given in the input data file; the only constraint is that the boundary of the entire structure must be made compatible with those outlined above.

With finite element simulations of field problems, it is efficient to arrange the mesh to be finer where the magnitudes of the field and/or its gradient are larger. Hence, adaptive FEM (AFEM) [119]-[122] may be preferred. Since the efficiency of AFEM is structure-dependent and not always superior as compared with the fixed-grid FEM (FGFEM), an alternative approach of using pre-graded grids (PGGFEM) is worth trying owing to the fact that in most cases the field patterns are known qualitatively *a priori*. In principle, arbitrary gradings can be incorporated in PGGFEM, but it is difficult to achieve without the user's intervention to the source code. As a compromise between convenience and efficiency, PGGFEM is performed on a region-by-region basis. Of course, some average is necessary at inter-region interfaces to produce a good mesh. To further improve the gradation and shapes of elements in a mesh, a smoothing procedure is incorporated.

As is known, the matrix equation resulting from FEM is quite sparse. It is desirable to make use of the sparsity so that both storage and computation time can be reduced a great deal. An obvious approach is to arrange a banded form of matrix by renumbering nodes in a mesh.

Finally, second-order triangular elements are generated on the basis of the first-order elements, and a novel renumbering scheme for the nodes of the higher-order elements is proposed.

4.2   Fundamentals

Some nomenclature commonly encountered in the text is now defined:

Domain A connected and closed subset of a plane and the largest area to be considered.

Region A closed subdomain throughout which the same material code and vertex node spacing is specified.

Interface The intersection of two regions.

Boundary A directed, closed and minimum subset of a domain or region such that every element in the domain or region falls either on or to the left of it.

Edge A straight line joining two nodes.

An edge of a region: a straight line joining two corner nodes of the region.

An edge of an element: a straight line joining two vertices of the element.

The types of region which can be embedded in a domain are shown in Figure 4.1, which are labelled as Types 1 to 5. Type 5 is restricted to be concentric, but Types 2 and 3 may be non-centrosymmetric provided that inner and outer boundaries do not touch. The edges of Types 1 to 3 have to be parallel with or perpendicular to the horizontal axis.

A domain can then be constructed by any combination of these types of region provided it is compatible. The compatibility requires that any edge or circular boundary of one region must be common to any adjacent region. For example, the structures in Figure 4.2 are all compatible whereas those in Figure 4.3 are incompatible. It is apparent that Figure 4.3 (a) and (b) can be made compatible by further dividing the structure, and Figure 4.3 (c) and (d) can be handled by treating them as entirely "inhomogeneous" problems.

For the above rather regular structures, the generation of evenly distributed nodes in each region is greatly simplified. It will become clear that the restrictions outlined here could have been easily relaxed in the automatic triangulation and smoothing procedures and have no effect on the node renumbering.

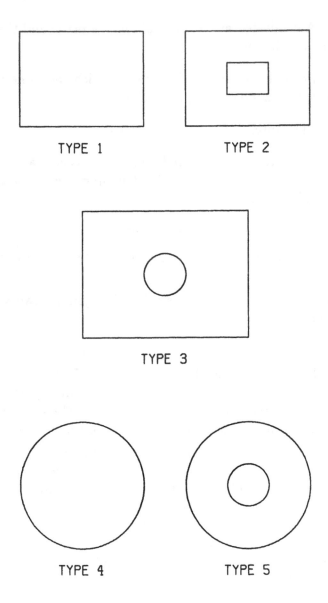

Figure 4.1: Types of regions which can be embedded in a domain, where the Types 2, 3 & 5 are hollow.

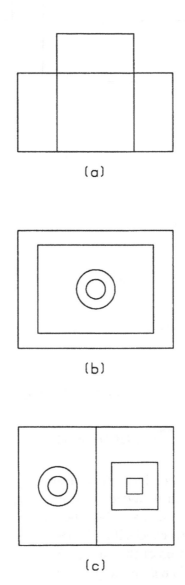

Figure 4.2: Examples of compatible structures.

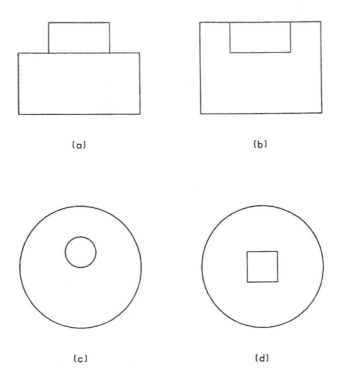

Figure 4.3: Examples of incompatible structures.

4.3 Node Generation

Vertex node generation has a great bearing on element generation. Poorly shaped elements can be avoided by ensuring that node positions cannot give rise to such elements. Also, the triangulation into elements can be carried out with few checks if node positions are optimised in some sense. Thus the modification of the Fukuda-Suhara method [123] made by Shaw and Pitchen [124] is a great step in approaching a more direct generating method than simply generating interior nodes randomly.

It should be mentioned that there are two kinds of poorly shaped elements: one kind are poorly shaped triangles which are acute-angled; the other kind are those elements whose three vertices are all on the exterior boundary of the domain. In [112] and [124], the first kind are avoided by homogeneous node distribution; the second kind, however, can obviously arise. In the following the former will be termed "poorly shaped triangles" and the latter "trivial elements". Trivial elements can be avoided in the element generation stage.

The vertex node generation scheme employed, similar to those in [112, 124], is as follows:

1. The domain is divided into a number of regions with different material code and/or vertex node spacing. Each region must match one of the five allowed types.

2. Nodes are generated along boundaries and interfaces of the different regions according to the required node spacing. The geometrical average is assumed along the interfaces of regions with different node spacings.

3. Interior nodes are generated in a region as follows:

    3.1 If the region belongs to Type 2, it is divided into four subregions of Type 1, as is illustrated in Figure 4.4.

    3.2 If the region belongs to Type 3, it is treated as Type 1 except that no nodes are generated within the inner circular region or at a distance of less than half of the node spacing from the circular boundary.

    3.3 If the region is of Type 4, it is taken as Type 5 with zero radius for the inner circular boundary.

    3.4 Interior nodes in a Type 1 region are generated in the following way:

    - Equally-spaced imaginary horizontal lines are drawn between y_{min} and y_{max} across the region. The spacing between the horizontal lines is about $\sqrt{3}/2$ of the desired node spacing, depending on rounding-off.

    - Equally-spaced nodes are generated along each horizontal line according to the desired node spacing and rounding-off, and are placed in such a manner on adjacent lines that nodes at the centre of the region will produce equilateral triangles automatically when there is no rounding-off. For the two horizontal lines adjacent to the exterior boundary, the first and the last node on the line are at half node

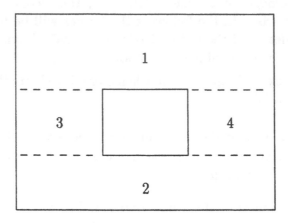

Figure 4.4: Decomposition of a Type 2 region.

spacing to their adjacent vertical edge in order to avoid "trivial elements" while keeping "good shape" there.

3.5 Generation of interior nodes in a Type 5 region is similar to that in a Type 1 region except that horizontal lines are replaced by concentric circles and y_{min} and y_{max} are replaced by r_{in} and r_{ex}. The "trivial elements", if any, are avoided in the triangulation stage.

3.6 If more subregions of Type 1 are to be completed, step 3.4 is repeated.

4. If more regions are to be completed, step 3 is repeated.

Here y_{min} and y_{max} are the minimum and maximum vertical coordinates of the region, and r_{in} and r_{ex} the radii of the interior and exterior boundaries of the region, respectively.

Indeed, the "poorly shaped triangles" are avoided with the present node generation scheme by incorporating the triangulation techniques described in the next section. As an example, the generated nodes of a two-region structure are illustrated in Figure 4.5.

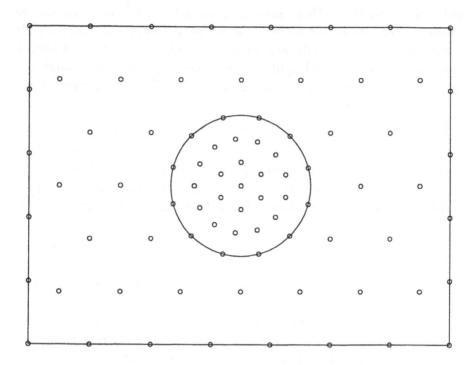

Figure 4.5: Illustration of generated nodes for a two-region structure.

4.4 Automatic Triangulation and Smoothing

Nodes are generated rather evenly in each region as described in the previous section. Consequently, less control need be performed in triangulating the mesh. A good strategy for data management can also improve the efficiency.

The algorithm of triangulation developed by Lo [112] is novel and efficient. It is much faster than that of Cavendish [113] and Shaw-Pitchen [124] since consideration of nodes on the generation front and in its interior instead of in the original region renders unnecessary the check to see if the new triangle overlaps any previously generated triangles or any fixed interface segments. This not only saves much computation time but greatly stabilised the triangulation process [112]. However, the criterion for the choice of a node C to complete a new element with base $\overrightarrow{AB}$ in

[112] is the minimum value of the norm $[AC^2 + CB^2]$. This criterion, as pointed out by the author himself, may not be sufficient to guarantee the best triangulations for regions having very irregular boundaries. It has been found that the criterion is not sufficient even for regions having regular boundaries but with quite different node spacings. Although the criterion of considering more than one node is available, this slows down the process a great deal.

It is the experience of the author that the maximum angle $\angle ACB$ is a better criterion for the choice of the node C with the given base $\overrightarrow{AB}$ in the sense of best triangulation and is much faster than the algorithm of considering more than one node as adopted in [112]. The maximum angle criterion has been used by Shaw-Pitchen [124] and works fairly well, as is shown in their results.

The automatic triangulation algorithm adopted here can be considered as the combination and modification of the S.H. Lo [112] and the Shaw-Pitchen [124] methods. As well, some new features are exploited. The new algorithm is expected to be more reliable in obtaining the best triangulation as compared with that of S.H. Lo when one node is considered for a new element [112], and faster than the Shaw-Pitchen method and the Lo method when more than one node is considered for a new element. The modified algorithm is stated as follows:

1. Form a list of triangular bases as the generation front for which apex nodes must be found. Initially these will be the collection of boundary segments for the region.

2. Select a base $\overrightarrow{AB}$ from the front and search for a node C which lies to the left of $\overrightarrow{AB}$ from the interior nodes or from the front itself such that $\angle ACB$ is a maximum, provided that node C is not on the boundary of the region when both nodes A and B are on the boundary.

3. Complete $\triangle ABC$ and update the generation front.

4. If the generation front is not empty, return to 2.

It is apparent that the above triangulation scheme is distinguished from that of Shaw and Pitchen by a rather different algorithm for triangulation and differs from that of Lo in the criterion of the choice of a node C. It is a practical trade-off between best triangulation and computation time as well as the effort of programming. Trivial elements are totally banned from the formation, which is an added feature.

Besides the systematic arrangement of the input data, similar to that of Lo, another step made towards a systematic approach is that nodes of each element are stored in a counter-clockwise order. This makes the finite-element programming easier since the procedure for establishing whether the line segment $\overrightarrow{AB}$ is on an interface between different media reduces to checking whether the material code of the element formed by base $\overrightarrow{BA}$ is different from that of the element formed by base $\overrightarrow{AB}$, and a check to see if the line segment $\overrightarrow{AB}$ is on the boundary of the domain becomes simply a verification that there is no element formed by base $\overrightarrow{BA}$.

The mesh network based on the generated nodes shown in Figure 4.5 is plotted in Figure 4.6.

Though all the five types of regions which can be embedded in a domain are convex, the auto-triangulation algorithm adopted can be directly applied to nonconvex regions thanks to the formation of elements from *vector* bases. In some other strategies, however, a nonconvex region has to be decomposed to a number of convex subregions, *e.g.* [125, 126].

The triangular mesh network is then smoothed by shifting each interior node of each region to the centre of its surrounding polygon. This process of node relaxation is repeated until a certain convergence criterion is met. This method of smoothing suggested by Cavendish [113] and also adopted by Lo [112] is rather simple but very effective. Theoretically, the nodes at interfaces and boundaries have one degree of freedom and can also be relaxed; but they are excluded from relaxation because they are already evenly distributed. Another reason is that the relaxation of those nodes can degrade the network when the ratio of node spacings of the two adjacent regions is large. The smoothed mesh network of Figure 4.6 is shown in Figure 4.7.

4.5 Automatic Node Renumbering

A common feature of finite element analyses is the generation of a large number of algebraic equations describing the interaction between related nodes. Methods for the efficient solution of these equations require their

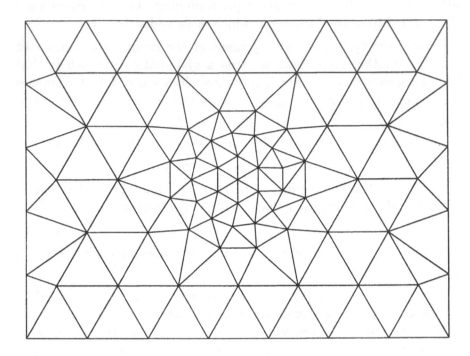

Figure 4.6: Mesh network resulting from the generated nodes in Figure 4.5.

storage in the most economical way possible. The efficiency of storage of the matrix depends on the solution method adopted as well as the properties of the operator (*e.g.* the self-adjointness). Even so, there is no doubt that the way in which the nodes are numbered can greatly affect both the bandwidth and the profile of the matrix for any given system. Thus, many strategies for automatic reordering of nodes have been proposed to reduce bandwidths and/or profiles. A comprehensive survey of these methods can be found in [127]. Perhaps the best known of these are the methods of Cuthill-McKee [128] (CM), reverse Cuthill-McKee [129] (RCM) and Gibbs-Poole-Stockmeyer [130] (GPS), which are very widely used and are generally fast and efficient. Other typical methods such as those described by Levy [131], Grooms [132], Puttonen [133], Akhras-Dhatt [134], George-Liu [135], Burgess-Lai [136, 137] and Konishi-Shiraishi-Taniguchi [138], *etc.* have their own particular philosophy and are worth considering. In the literature, the various methods have been tested with many different kinds of structures, but the algo-

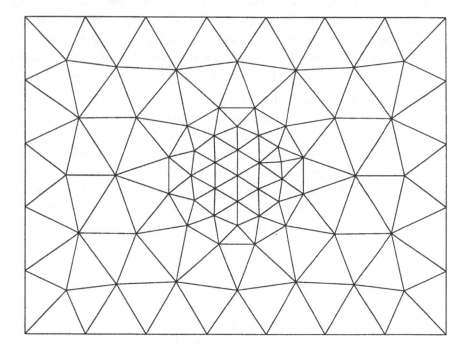

Figure 4.7: The smoothed mesh network of Figure 4.6.

rithms are either inefficient or complicated and also require a lot of human effort in programming. In view of this, the question may be asked whether there are any simple but rather efficient algorithms suited to a restricted class of structures. The node renumbering scheme suggested in this work is just such an algorithm. It has been developed for 2D mesh networks, especially structures with rectangular exterior boundaries, but an extension to 3D mesh networks is also possible. This new algorithm for minimising the bandwidth consists of two stages: the choice of a starting node and the node renumbering.

The criterion for choosing a starting node can be either the node with lowest connectivity (CM method) or the endpoint of a pseudo-diameter (GPS method). With evenly distributed nodes, generally speaking the node with lowest connectivity and the endpoint of a pseudo-diameter should always fall on the boundary of the domain. Also, testing the connectivity of boundary nodes is much easier than finding a pseudo-diameter. Thus, in the new algorithm the potential candidates are chosen from only the collection of the boundary nodes according to the connectivity and the final choice is made for the one demonstrating best performance. For simply-connected domains, the starting node selected from the above criterion has a high possibility of being an endpoint of an optimal pseudo-diameter for regular structures, and the potential candidates will not be many for structures with rectangular boundaries.

Once a starting node is determined, the original structure is converted to a *level structure*, as shown in Figure 4.8. The new scheme of node renumbering is distinguished from that of CM and GPS due to the fact that, within a level, nodes are labelled counter-clockwise[1] rather than in the order of increasing connectivity. Therefore, it is expected to be faster than the CM and GPS methods since, within a level, it is much easier to establish a counter-clockwise sequence of nodes than to find out the connectivity of each node and then compare the bandwidth for all the alternatives among nodes with the same connectivity. Also, it can be easily applied manually if desired. An illustrative example is shown in Figure 4.9.

The algorithm adopted is by no means optimal even in the case of rectangular and/or circular structures with simplex elements. Actually, there may be no optimal algorithm at all, even for such simple structures, except trying all the possibilities exhaustively, which is far from being practical for large problems. However, it has worked quite satisfactorily for all the waveguide structures tested.

Of course, its effectiveness should be verified by comparing the results from the present algorithm with those reported in the literature. Unfortunately, such comparisons are frustrating since most of the examples in the publications available either give insufficient information on the structures or are too complicated to prepare a data file. Also none of the examples is suitable for application in this work and only a few examples are networks of simplex triangular elements. Also several mistakes have been found in the reported results with renumbered node sequences given. All this make comparisons almost meaningless, difficult and time consuming. Nevertheless, a comparison was made for some simple examples in the literature with known minimised bandwidth and possibly renumbered node sequences, *e.g.* Figures 5, 7 and 8 in [132] and Figure 8 in [139]. Our results are very favourable and comparable to the best.

The above examples compared are rather simple having only a few

---

[1]The counter-clockwise choice is arbitrary, but once made, must be kept for all levels.

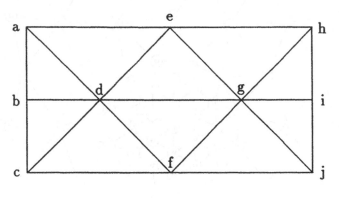

(a) Original structure.

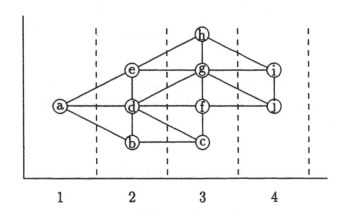

(b) Level structure.

Figure 4.8: An example of level structure.

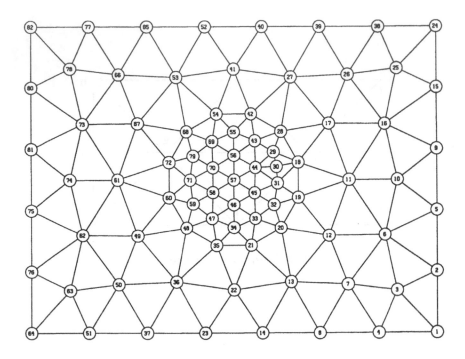

Figure 4.9: The node sequence of Figure 4.5 from the present node renumbering strategy. (Final bandwidth: 16, No. of nodes: 82, No. of elements: 138).

dozen nodes and the FEM systems resulting from them can be solved immediately for most practical problems by a modern computer even without node renumbering. The renumbering approach can only be appreciated for large problems. As a demonstration of the effectiveness of the proposed algorithm and the importance of node renumbering, two relatively large examples are given in Figure 4.10 with the reordered node sequences shown therein, where the final bandwidth (actually, semi-bandwidth) is defined as the maximum difference between any two related nodes plus one to account for the diagonal term. "Two related nodes" has the same meaning as "two connected nodes" *iff* all the elements in the mesh network are simplex.

Although the proposed algorithm is mainly for structures compatible with those outlined previously, it can be applied to any structure including 3D mesh networks. As for unusual structures with very irregu-

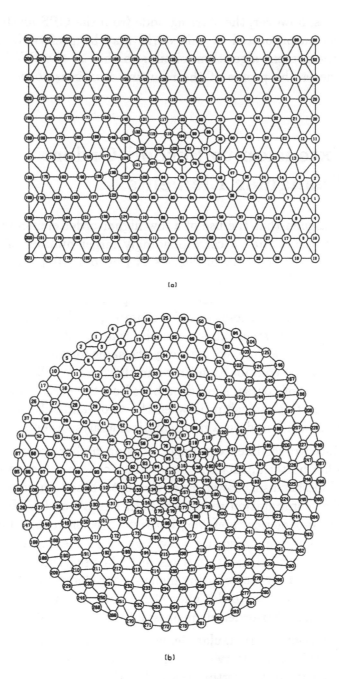

(a)

(b)

Figure 4.10: A demonstration of the effectiveness of the present node renumbering algorithm for relatively large problems. (a) Final bandwidth: 18, No. of nodes: 209, No. of elements: 364. (b) Final bandwidth: 23, No. of nodes: 286, No. of elements: 517.

lar boundaries, however, the starting node from the GPS method, which
has not been incorporated in this work, may be preferred. Also, the
reverse of the present algorithm, similar to that of the RCM, is worth
trying whenever the profile of the resulting matrix is to be minimised
[140].

4.6   Extension to Second-Order Triangular Elements

It has been argued that the use of a few higher-order elements produces
much better results than a correspondingly larger number of low-order
ones with the same number of degrees of freedom [102]. However, one
should bear in mind that the same number of degrees of freedom does
not mean the same cost, *i.e.* the same computation time and storage requirements. For banded solutions, for example, the bandwidth resulting
from second-order elements almost doubles that resulting from first-order
elements, as a general rule, with exactly the same number of nodes. Consequently, both the computation time and the storage will be doubled. A
fair comparison, of course, should be based on the cost. Unfortunately,
there is no exact relationship between the two bandwidths. Therefore,
such a comparison based on a mathematical analysis seems impractical.

The incontrovertible advantage of using higher-order elements is
that they offer better ultimate accuracy than can be achieved by the refinement of meshes using lower-order elements, based on error analyses,
provided that computation resources (both storage and CPU time) are
not limited. In practice, however, both storage and computation time
are usually restricted. Also, for very high order elements, say $p \geq 3$,
the derivation of the element matrices in FEM programming is very involved. Another drawback of using higher-order elements is that it is not
as flexible as using lower-order ones at irregular boundaries. The most
commonly used higher-order element is only of second order.

Given first-order triangular elements, the generation of the second-
order elements is straightforward: simply add another node halfway between any two connected vertex nodes. The difficult part involved is how
to reorder the nodes *effectively* in the new mesh network. In principle,
all the renumbering strategies that can be applied to first-order element
networks can also be applied to second-order networks. Nevertheless a
satisfactory bandwidth minimisation is not guaranteed. Being unable to

find a resequencing algorithm that is specialised to or has proven effective for second-order triangular elements, the author tried an easy and efficient method which has proven to be very effective. Instead of directly applying the node renumbering algorithm stated in the previous section to the second-order element network, each second-order element is conceptually divided into four first-order elements as is shown in Figure 4.11, and the renumbering is performed on the first-order mesh in the same way as before. This *indirect* strategy does yield much better results than the direct renumbering. A comparison of these two strategies for a typical example is shown in Figure 4.12.

The indirect technique used here can be easily extended to any other high-order and/or hybrid elements as well as to 3D mesh networks by the conceptual conversion of each non-simplex element to a number of simplex elements. The widely used CM and GPS methods can be improved by utilising this technique for non-simplex elements. Further examples of its use are given in Chapter 6.

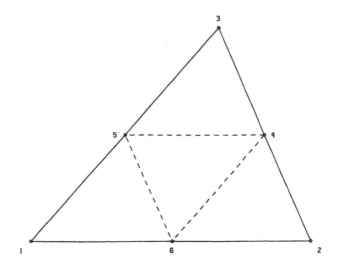

Figure 4.11: Decomposition of a second-order element into four first-order elements.

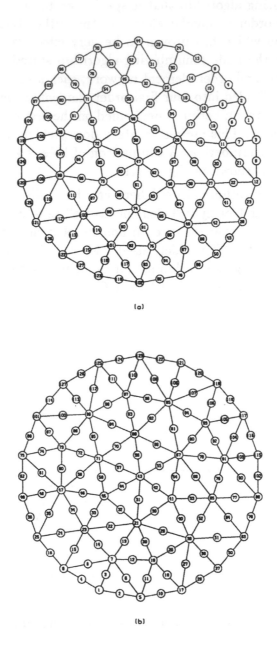

Figure 4.12: A comparison of the effectiveness of the direct and indirect renumbering strategies for second-order elements (127 nodes, 54 second-order elements). (a) Direct (final bandwidth: 46). (b) Indirect (final bandwidth: 31).

Chapter 5

Solution of Nonlinear Generalised Eigenvalue Problems

5.1 Introduction

In a nonlinear situation the matrices in the global matrix equation (Equation 3.22) derived in Chapter 3 are implicit functions of the unknown electromagnetic field. Because the nonlinear term is small and the matrices are sparse, these nonlinear algebraic equations can be solved more efficiently as essentially linear equations by using a simple nonlinear iteration scheme to account for the nonlinearity than as nonlinear equations requiring Newton-like methods [141]-[143].

The nonlinear algebraic matrix equation (Equation 3.22) has nontrivial solutions consisting of some discrete eigenvalues of λ or $\bar{\beta}$ and corresponding eigenvectors[1] x and thus can be formally called an eigenvalue problem.

As our solution procedure strongly relies on linear eigenvalue problems, we need to concern ourselves with the efficient solution of linear generalised eigenvalue matrix equations. The choice of an efficient solution method for a given problem depends on many aspects, such as size, symmetry, positive definiteness, sparsity pattern, available RAM,

---

[1]In Chapter 3, α was used for such eigenvectors because x had been used for another purpose. In accordance with most mathematical literature, α will be replaced by x from here on.

the number of required eigenvalues and eigenvectors, *etc.* By and large iterative methods are efficient for 3D problems and direct methods are most suitable for 1D problems. Our problem, as it happens, should be considered as 2D here. Also, the efficient solution of sparse algebraic systems is at present an extremely active area of research in both the engineering and mathematics/computer science communities and many well-known existing solution methods are being improved all the time.

An algorithm is useful only when the error resulting from it is within an acceptable range. There are mainly two sources of errors in FEM computations: *approximation* (due to the discretisation of the structure and the shape functions of the elements) and *round-off* (due to computer word-length). At this stage one can do nothing about the approximation error; however, the round-off error may be minimised by choosing a suitable matrix solution algorithm. Like estimating the approximation error, it is possible to estimate the round-off error of a solution method qualitatively, but it seems impossible to do it quantitatively. It is impractical to establish a quantitative *a priori* error estimate, but a quantitative *a posteriori* error estimate of a particular solution is always handy; simply evaluate the residual upon obtaining the solution. This approach will be adopted in the present work.

The validity of a solution strategy may be verified in different ways. The most desirable way is to prove it mathematically, which is usually done by mathematicians. From an engineering point of view, more often than not the verification is done in an *a posteriori* sense by substituting the solution obtained into the original differential equations and evaluating the residuals, whereby at least C^1-elements are required in our case and thus the approach is rather restrictive. Alternatively the solution obtained could be compared with the exact one for a simple example having an analytical solution. Unfortunately, no known nonlinear waveguiding structures of finite geometrical extent can be solved analytically.

The algorithm for solving nonlinear waveguiding problems in this work can be justified in the following way. Firstly, a comparison is made with linear problems of known analytical solutions as our eigenvalue problem is a linear one at each nonlinear iteration step. Then a check is performed to ensure that the nonlinear iterative solution method converges properly and the error accumulated during the nonlinear iteration is within an acceptable range by evaluating the residuals of the nonlinear algebraic equations for each eigenpair obtained.

A simple nonlinear iteration scheme usually gives slow convergence.

Therefore, any acceleration technique which can speed up the process is an advantage. Though many other solution methods [144]-[152] are available and by and large converge 'faster' than a simple iteration method, it is misleading to compare different iteration methods by the number of iterations required to achieve certain accuracy. The merits of a solution method should be judged by proper weighting of the computation time and the storage required apart from the human effort involved. Of course, one has to admit that some complicated methods may be more reliable for problems of large nonlinearity. To return to our problem, the matrices are rather large and of banded structure; and the nonlinearity is relatively small. Thus, it is appropriate to adopt a simple nonlinear iteration scheme with a suitable acceleration technique whereby the sparsity pattern of the matrices is not destroyed, as is done in [25, 153].

5.2 The Generalised Eigenvalue Problem and the Nonlinear Iteration Procedure

In analysing nonlinear optical waveguides, the most useful dispersion relation is the $\beta/k_0 \sim P$ curve with k_0 and dimensions of the structure given. Therefore, it is preferred that the coordinates are normalised with respect to k_0 rather than β. In normalising the structure with respect to k_0, however, the eigenvalue problem has a non-standard form, which needs a special technique to solve, as mentioned in Chapter 2. While the iterative procedure for the solution of the generalised nonlinear eigenvalue problem of standard form can be found in [25], we only consider the solution of the non-standard nonlinear eigenvalue problem in the following.

By using the transcription $\frac{\partial}{\partial \bar{z}} \rightarrow -j\bar{\beta}$, the global matrix equation (Equation 3.22) can be written explicitly in terms of the actual eigenvalue $\bar{\beta} = \beta/k_0$ in the form:

$$[\bar{\beta}^2 A + \bar{\beta} B + C]\, \boldsymbol{x} = \boldsymbol{0} \tag{5.1}$$

when the normalisation factor $\rho = k_0$ and consequently $\lambda = 1$. This is a quadratic eigenvalue problem and is solvable with some rather expensive methods [77, 154]. Alternatively, Equation 3.22 can be expressed as a generalised eigenvalue problem of the standard form and the same size:

$$R(\bar{\beta})\, \boldsymbol{x} = \lambda T\, \boldsymbol{x} \tag{5.2}$$

where

$$R \triangleq S + pU \tag{5.3}$$

and solved iteratively by modifying $\bar{\beta}$ (letting $\bar{\beta}_{i+1} = \bar{\beta}_i/\sqrt{\lambda_i}$, where i is the iteration index) till $\lambda \to 1$. Here, λ is taken as the eigenvalue only formally, and the actual eigenvalue is $\bar{\beta}$. It can be shown that the spectra of $\bar{\beta}$ and λ are reversed with respect to one another by using the fact that the fundamental mode has the largest propagation constant for given frequency, and has the lowest frequency for given propagation constant.

Though this iterative approach is not recommended for linear problems as far as the computation time is concerned, it has been shown to be an effective method for nonlinear problems if $\bar{\beta}$ and $\hat{\epsilon}_r$ are modified simultaneously.

The generalised nonlinear eigenvalue problem in Equation 5.2 can be solved self-consistently by the following nonlinear iteration procedure:

a) specify the guided power P and calculate R and T (assuming no nonlinearity);

b) solve the linear eigenvalue problem in Equation 5.2 for λ and $\boldsymbol{x}$;

c) optionally accelerate or decelerate the nonlinear iteration process to get a modified λ and $\boldsymbol{x}$, and then scale $\boldsymbol{x}$ according to the given power P;

d) go to e) if the eigenpair does not converge within the desired criterion, and stop otherwise;

e) modify $\bar{\beta}$ and $\hat{\epsilon}_r$ to get new R and T;

f) repeat b) to d) to obtain modified λ and $\boldsymbol{x}$.

Note that the acceleration or deceleration should only be performed after the second pass of nonlinear iteration. If linear iterative methods are employed in stage b), the eigenpair found from the previous pass of *nonlinear* iteration is used as the initial value for the current *linear* iteration, which may already have several significant digits correct. The first time that stage b) is reached, the initial value used is arbitrary but nontrivial.

The above iterative approach is more or less the generalisation of the work in [25]. Both are based on seeking the effective index β/k_0 for given guided power P with the nonlinear iteration starting from a linear solution, which we call the conventional solution technique. In Chapter 6, we will develop a non-conventional solution technique, that is, seeking P for given β/k_0, as the need arises. Also, we will show how to modify the conventional solution technique to simulate a certain class of phenomena, which otherwise would be impossible.

5.3   Solution Methods

Due to the limited computational resources available, in the following we confine ourselves to the restricted anisotropy of the media as outlined in § 3.3, where the resulting global matrices are real symmetric.

As discussed previously, the efficiency of an algorithm depends on the detailed properties of the problem to be solved. Further examination of our problem is needed prior to selecting the solution method. The general properties of our linearised algebraic eigenvalue equation are summarised in Table 5.1. Apart from those factors under consideration for linear problems, other features arise from the iterative nature of the solution method used for our nonlinear problems. For a linear algebraic eigenvalue problem a few eigenpairs at one end of the spectrum can be computed simultaneously by *simultaneous vector iteration* [155] if desired; however for a nonlinear problem eigenpairs have to be found individually whenever a nonlinear iteration scheme is employed. Also, at each nonlinear iteration step, only one eigenpair has a physical meaning (except in the first step when the nonlinearity is set to zero); thus one must exercise care in computing the physical modes as the mode index of a physical mode might be swapped with that of an unphysical solution during the nonlinear iteration, particularly for higher-order modes.

Integrating the above properties, we believe that the *successive overrelaxation* (SOR) [156] and *vector iteration* (VI) [155, 157] methods are two of the most efficient methods for the computation of the fundamental mode for our rather large problems resulting from quasi-3D structures where the sparsity pattern can be utilised. As for higher-order modes, the *bisection and inverse iteration with shift* (BIIS) [154, 158, 159] method is the most appropriate for finding an individual eigenpair (the physical one) without being bothered by other lower-order (unphysical) ones and is thus efficient both computationally and in storage compared

PROPERTY	VALUE	COMMENTS
Number of degrees of freedom (NDF)	~ 1000	Approximately 1000 unknowns corresponding to about 330 nodes
Symmetry	Symmetric	The matrices and unknowns are also real
Definiteness	Positive definite	Both R and T matrices
Sparsity	Banded	The semi-bandwidth is around 8% of NDF when NDF is about 1000 for first-order elements and is about 15% for second-order elements
Eigenvalue and/or eigenvector required	Both	Only one or two lowest modes
Computational resources (CPU time and storage)	Both are limited	CPU time is more crucial than storage

Table 5.1: General properties of linearised algebraic eigenvalue problems in the present work.

with the *Lanczos* [77, 155, 160, 161], the *deflation* [155, 161], the *purification* [155] and the *block/subspace/simultaneous iteration* [154, 155, 161] methods.

The methods selected for solving linear generalised eigenvalue problems are described in the following.

5.3.1 Successive overrelaxation and Rayleigh quotient

The SOR method is one of the oldest and most successful linear iterative methods for sparse system computations. In most of the textbooks, the SOR method is used to compute the deterministic problem:

$$A\boldsymbol{x} = \boldsymbol{b}. \tag{5.4}$$

The iteration is carried out in the form [156]

$$D\boldsymbol{x}_{i+1} = \omega_r(C_L\boldsymbol{x}_{i+1} + C_U\boldsymbol{x}_i + \boldsymbol{b}) + (1 - \omega_r)D\boldsymbol{x}_i \tag{5.5}$$

with a starting vector $\boldsymbol{x}_0$, where ω_r is a real number known as the *relaxation factor* or *relaxation parameter*, and D, $-C_L$ and $-C_U$ are, respectively, the diagonal, strictly lower triangular, and strictly upper triangular parts of A.

If A is symmetric and positive (or negative) definite, SOR methods converge to the solution for any fixed value of ω_r in the range $0 < \omega_r < 2$ [157]. Only overrelaxation ($\omega_r > 1$) is of importance for fast convergence, after which the method is named. When ω_r is unity, the SOR method reduces to the Gauss-Seidel method.

The SOR method can be directly adapted to the generalised real symmetric eigenvalue problem

$$K\boldsymbol{x} = \lambda M\boldsymbol{x} \tag{5.6}$$

to solve for the eigenpair corresponding to the smallest eigenvalue as was done in [162]-[165]. In this connection, one simply sets

$$A = K - \lambda M \tag{5.7}$$

and

$$b = 0 \qquad (5.8)$$

in Equation 5.4. After each step of iteration the eigenvalue λ is estimated by the *Rayleigh quotient* (RQ) [77] so that

$$\lambda \rightarrow \lambda_i = \frac{x_i^T K x_i}{x_i^T M x_i} \qquad (5.9)$$

which is used for the next step of iteration until some convergence criterion is met; and the eigenvector x_i is normalised with respect to the quadratic form for M so that

$$x_i \rightarrow \bar{x}_i = \frac{x_i}{(x_i^T M x_i)^{\frac{1}{2}}} \qquad (5.10)$$

after each step of iteration to avoid overflow on a computer of limited word-length. The SOR-RQ described here has a global convergence and converges to the eigenpair corresponding to the smallest eigenvalue.

As discussed above, the convergence of the SOR method is guaranteed only for a real symmetric definite matrix A whereas the matrix "A" defined in Equation 5.7 is becoming singular as the estimated eigenvalue approaches its exact value. Therefore, there might be a potential danger of lack of convergence in the scheme. Also, in using SOR, it is quite important to choose a good value of ω_r as the rate of convergence of the SOR iteration is substantially reduced if ω_r is not near its optimum value. This is one of the great drawbacks of SOR [161]. Luckily, in our computations satisfactory convergence has always been achieved for $1.2 \leq \omega_r \leq 1.7$, where the convergence seems not to be too sensitive to the relaxation parameter within the above range. However, when the choice of a good value of ω_r becomes a problem for rapid convergence, the *symmetric successive overrelaxation* (SSOR) method and the *adaptive successive overrelaxation* (ASOR) method are worth trying. In this connection, we refer particularly to the excellent book by Hageman and Young [156] and the references therein.

5.3.2   Vector iteration

An alternative way of solving the fundamental mode is the VI method motivated by the *power iteration* (PI) method [157, 161], which is also very efficient. The VI method can be classified as either *forward iteration* (FI) or *inverse iteration* (II), where the former converges to the eigenvector of the largest eigenvalue and the latter converges to that of the smallest eigenvalue [157]. In solving Equation 5.6 for the fundamental mode, we attempt to find the eigenpair corresponding to the smallest eigenvalue. Therefore, only II will be considered here and briefly described in the following; more details can be found in [155, 157, 161].

The eigenvector of the smallest eigenvalue in Equation 5.6 can be computed iteratively by solving

$$K\boldsymbol{x}_{i+1} = M\boldsymbol{x}_i \qquad (5.11)$$

for $\boldsymbol{x}_{i+1}$; and the smallest eigenvalue is estimated by the RQ defined in Equation 5.9 after each step of the iteration. Then the approximated eigenvector is normalised according to some rule, *e.g.* Equation 5.10; and the normalised eigenvector is substituted into the right-hand side of Equation 5.11 for the next iteration. This process is repeated until the desired convergence is achieved. The II-RQ method yields quadratic convergence and converges to the eigenpair of smallest eigenvalue globally. The II-RQ method may be accelerated by switching the II to the *inverse iteration with shift* (IIS) (see next subsection) after a certain stage with some risk of converging to the wrong eigenvalue, as the IIS-RQ method has cubic convergence near its solution [161].

Equation 5.11 is a deterministic problem and can obviously be solved efficiently by the SOR family as described previously provided that the matrix K is real symmetric positive definite. Note that the iteration matrix K does not change during the II and thus can be factorised once and for all by *Cholesky decomposition* [77, 157] if K is real symmetric (or complex hermitian) positive definite, or by LDL^T *decomposition* [155] for K being real symmetric (or complex hermitian) but not necessarily positive definite, or by *LU decomposition* [77, 155, 157] for general (real or complex) K; then Equation 5.11 is solved by *forward* and *backward substitution* [157], which is also very fast except that it requires some extra storage as compared with the SOR. All of the above three decomposition methods do not destroy the banded structure of the sparsity pattern. Another suitable method for solving sparse deterministic matrix equations such as Equation 5.11, which has been receiving more and more attention in recent years, is the *preconditioned conjugate gradient* (PCG) family. Examples of PCG methods are given in [152, 161] and [166]-[172].

5.3.3   Bisection and inverse iteration with shift

The eigensolutions of Equation 5.6 are related to those of the shifted eigenvalue problem

$$(K - \bar{\lambda}M)y = \nu My \tag{5.12}$$

by

$$\nu_i = \lambda_i - \bar{\lambda} \tag{5.13}$$

$$y_i = x_i \tag{5.14}$$

where the subscript i is the modal index and $\bar{\lambda}$ is a given real number.

If Equation 5.12 is solved by the II method, the eigenvector found is the eigenvector of Equation 5.6 corresponding to the eigenvalue closest to $\bar{\lambda}$. This property is very useful, and enables us to find an individual eigenvector as long as we know a good approximation to its eigenvalue. Now the problem is how to estimate a desired eigenvalue of Equation 5.6. To answer this question, one introduces the *Sturm sequence property* [157].

The principal minors $p_r(\bar{\lambda}) \triangleq det\ (K_r - \bar{\lambda}M_r)\ (r = 0, 1, \cdots, q)$ of the symmetric matrix $(K - \bar{\lambda}M)$ with M being positive definite form a Sturm sequence provided that the corresponding matrix equation has no degenerate eigenvalues, where $p_0(\bar{\lambda}) \equiv 1$, K_r and M_r denote the principal rth-order submatrices of K and M, respectively. Consequently the number of eigenvalues strictly greater than $\bar{\lambda}$ is equal to the number of agreements in sign of this sequence provided that $p_r(\bar{\lambda})$ is taken to have the opposite sign to that of $p_{r-1}(\bar{\lambda})$ if $p_r(\bar{\lambda}) = 0$ [77]; or equivalently, the number of eigenvalues strictly less than $\bar{\lambda}$ is equal to the number of changes in sign of this sequence provided that $p_r(\bar{\lambda})$ is taken to have the same sign as that of $p_{r-1}(\bar{\lambda})$ if $p_r(\bar{\lambda}) = 0$ [155]. By using this property the kth eigenvalue can be isolated by repeated bisection and counting of the number of sign changes/agreements of the sequence.

The BIIS method has been explored by many researchers and now is a rather standard procedure for finding individual eigenpairs of a generalised symmetric eigenvalue problem. As for how to perform the bisection efficiently, how to accelerate the BIIS process and how to apply BIIS to degenerate eigenvalue problems, we refer to [77, 155, 157] and [173]-[175] for brevity. In the following are briefly described the methodology selected and certain modifications made to the existing strategies to improve the efficiency by making use of the special features of small bandwidth and single eigenvalue extraction in our problems.

In this work the principal minors are evaluated by *Gaussian elimination* with *partial pivoting* [155] to take advantage of the banded structure. With partial pivoting, however, the author has found that the Sturm sequence count may give misleading results whenever one or more of the pivots used are too small, which usually happens as the test eigenvalue is brought close to one of the exact eigenvalues and the separation between that eigenvalue and its neighbours is relatively small. These situations can be detected easily by comparing the number of sign changes/agreements when using the test eigenvalue with those when using its upper and lower bounds within the program. If this happens during the eigenvalue isolation stage, then the test eigenvalue is shifted by a certain mount, *say* 10% of the range between its bounds, from the current value and the Sturm sequence counting is restarted. The same remedy is also applied to the case of two consecutive principal minors happening to be *zero* where the Sturm sequence property does not hold. When the desired eigenvalue is well isolated, the bisection is replaced by the linear interpolation method, which not only offers faster convergence than the bisection method as claimed by others [155, 173] but also stabilises the process a great deal when the test eigenvalue approaches its true value. The usual equal bisection $(0 - 0.5 - 1)$ is replaced by the *golden section* $(0 - 0.618 - 1)$ method in the present work, which may be advantageous in a probabilistic sense. Another modification made to improve the efficiency is that instead of performing the sign counting on completion of the Gaussian elimination for the whole matrix in the conventional way, the counting is done after obtaining the sign of each principal minor. If the number of sign changes is greater than or equal to the index of the desired eigenvalue (in ascending order), further elimination would be useless as only the single physical eigenvalue is required, and the golden section is restarted with new bounds until the desired eigenvalue is isolated. The saving in computation time is considerable with this strategy whenever many eigenvalues are contained within the interval of the initial bounds of the desired eigenvalue, as is usually the case. It should be mentioned that the parabolic interpolation [176] is worth trying though the linear interpolation method is adopted in this work for the time being.

To complete this subsection, we would like to draw the reader's attention to the convergence of the BIIS method for nonlinear eigenvalue problems: if the index of the desired physical mode were exchanged with that of one unphysical mode or another during the nonlinear iteration, the BIIS method would fail to give convergence; and, whenever that is the case, under-correction is required during each step of the nonlinear iteration procedure as suggested in [102].

5.4   A Posteriori Error Estimates

Even if the nonlinear algebraic equations can be solved by the prescribed methods, probably we have to be armed with a certain metric or norm to determine how well an approximate solution converges to the exact one, which is usually done by establishing some error bounds. For nonlinear problems the error-bound approach is not as tractable as for linear problems; thus we have to adopt an alternative approach which is to estimate the residuals upon obtaining a solution.

For the generalised eigenvalue problem in Equation 5.6, we introduce a residual vector r defined by

$$r \triangleq K\tilde{x} - \tilde{\lambda}M\tilde{x} \qquad (5.15)$$

where $(\tilde{\lambda}, \tilde{x})$ is an approximate eigenpair. Then the "length" of the residual may be evaluated by using the norm associated with an appropriate inner product:

$$
\begin{aligned}
\|r\| &\triangleq \; <r,r>_q^{\frac{1}{2}} \\
&= \left(<K\tilde{x}, K\tilde{x}>_q + \tilde{\lambda}^2 <M\tilde{x}, M\tilde{x}>_q - 2\tilde{\lambda} <K\tilde{x}, M\tilde{x}>_q \right)^{\frac{1}{2}}
\end{aligned}
$$

$$(5.16)$$

where the real symmetric property of K and M has been utilised.

The above error estimate does not make much sense as its magnitude depends on the normalisation of $\tilde{x}$. To establish a useful error estimate, we introduce a scalar quantity r_r, called the *relative residual*, defined by

$$
\begin{aligned}
r_r &\triangleq \frac{\|r\|}{(<K\tilde{x}, K\tilde{x}>_q + \tilde{\lambda}^2 <M\tilde{x}, M\tilde{x}>_q)^{\frac{1}{2}}} \\
&= \left(1 - \frac{2\tilde{\lambda} <K\tilde{x}, M\tilde{x}>_q}{<K\tilde{x}, K\tilde{x}>_q + \tilde{\lambda}^2 <M\tilde{x}, M\tilde{x}>_q} \right)^{\frac{1}{2}}.
\end{aligned}
$$

$$(5.17)$$

It is apparent that r_r belongs to the set $[0,1]$ and is independent of the scaling of $\tilde{x}$; and r_r vanishes as the approximate eigenpair approaches the exact one.

Up till now we have not defined the inner product $< \cdot, \cdot >_q$ in R^q though the error estimate has been established. To be specific, we choose

$$< x, y >_q \triangleq x^T y, \qquad \forall x, y \in R^q \qquad (5.18)$$

in this work.

In order to compare the results from this work with the exact ones of a classical linear example, we have to establish another error estimate. Let $\tilde{v}$ be an approximate mode computed from this work and let the corresponding exact mode be denoted by $\check{v}$ with the *sign* being consistent with that of $\tilde{v}$ so that $< \tilde{v}, \check{v} > \geq 0$. Following a procedure similar to the above, we define a scalar quantity $e_r(\tilde{v}, \check{v})$, called the *relative error*, by the following metric:

$$e_r(\tilde{v}, \check{v}) \triangleq (1 - < \tilde{v}, \check{v} >)^{\frac{1}{2}} \qquad (5.19)$$

subject to the normalisation

$$< \tilde{v}, \tilde{v} > = < \check{v}, \check{v} > \equiv 1. \qquad (5.20)$$

Here the inner product takes the form of Equation 2.6; and e_r also belongs to the set $[0, 1]$ and approaches *zero* as $\tilde{v}$ approaches $\check{v}$.

5.5 Application of the Package to a Classical Linear Example

As discussed in the introduction to the present chapter, our iterative solution method for nonlinear guided waves strongly relies on the solution of linear waveguiding problems, and thus it is of great importance to test the methodology against linear structures. The linear example selected is a dielectric-loaded waveguide, as shown in Figure 5.1. This classical structure has analytical solutions [177] and has been adopted by many authors to verify their methods [14, 16, 21, 50, 54, 78], [82]-[84], [87, 98], [178]-[180]. To perform numerical analyses, the parameters are chosen as follows: $t = d = a$, $n_a = 1.5$, $n_b = 1.0$ and the relative permeabilities in both regions are assumed to be unity.

Our eigenpair extraction algorithms for both the fundamental and the higher-order modes can be tested by computing the first two lowest modes of the whole structure. (Actually, for this simple case, only

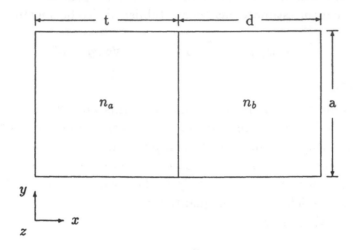

Figure 5.1: A dielectric-loaded waveguide, where n_a and n_b denote the refractive indices in each region and the perfectly conducting boundary condition is assumed.

half the structure need be solved if symmetry is used.) The geometrical structure in Figure 5.1 is meshed into 219 first-order elements with 130 nodes, as shown in Figure 5.2, where the node spacings in the two regions is deliberately made slightly different. The semi-bandwidth produced by our node renumbering scheme in terms of nodes is 13; therefore the semi-bandwidth in the matrix equation is only 39 which represents a great saving in computation time and storage as compared with the full bandwidth of 390.

The formulation used for the present example is the E-vector version. For the fundamental mode the computation is performed by the SOR method with an overrelaxation factor of 1.7. For the second lowest mode the computation is done by the BIIS method and the II is achieved by LU decomposition plus the forward and backward substitution method. Note that Cholesky decomposition cannot be applied to the matrix resulting from *inverse iteration with shift* as the system may not be positive definite. The dispersion curves of the first two lowest modes computed by the package developed here are compared with the exact ones. The accuracy of the present solution algorithm is clearly illustrated in Figure 5.3, which shows no noticeable difference between the FEM solution and the exact one for the fundamental mode (LSE_{10}

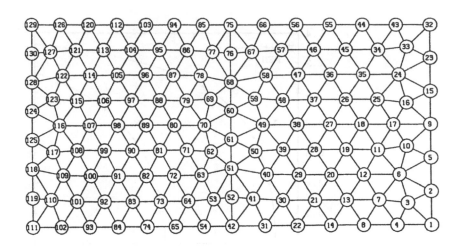

Figure 5.2: The mesh network of the dielectric-loaded waveguide structure with 130 nodes, 219 first-order elements and maximum node index difference between any two related nodes of 12.

or TE_{10}) with only 130 nodes, and a slight difference for the second lowest mode (LSM_{11}) which needs a finer mesh to improve the accuracy. The percentage errors of the computed normalised eigenfrequency $k_0 a$, defined by

$$\{\cdot\}_{error} = \frac{\{\cdot\}_{fem} - \{\cdot\}_{exact}}{\{\cdot\}_{exact}} \times 100\%, \tag{5.21}$$

are plotted in Figure 5.4, showing that the higher the frequency and thus the effective index, the better the accuracy, especially for the higher-order mode. This tendency may be advantageous in computing guided modes of optical structures, where effective indices are always greater than one.

The eigenvector computed for the fundamental mode at effective index $\beta/k_0 \approx 1.20$ is displayed in Figure 5.5, where the E_x and E_z components are several orders of magnitude less than the dominant E_y component but not identically equal to the theoretical *zero* due to inevitable round-off errors in computation. To compare the FEM solution with the exact one, the magnitudes of the electric fields from both the FEM and the analytical solutions are plotted in Figure 5.6, where the magnitude of the electric field at a point (x, y) is defined by the norm

$$\|E\|_2 \stackrel{\triangle}{=} \sqrt{|E_x|^2 + |E_y|^2 + |E_z|^2}. \tag{5.22}$$

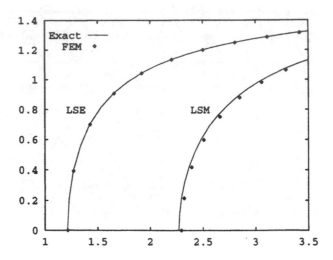

Figure 5.3:   Dispersion curves of the first two lowest modes of the dielectric-loaded waveguide: β/k_0 vs $k_0 a$.

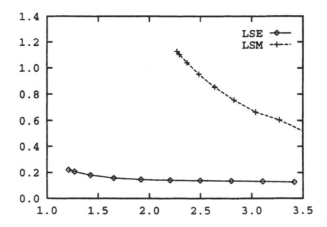

Figure 5.4: The percentage error of the computed normalised eigenfrequencies $k_0 a$ of the dielectric-loaded waveguide.

One can hardly notice the difference between the FEM solution and the analytical solution.

To obtain a quantitative *a posteriori* error estimate of the computed fundamental mode, the relative error defined in Equation 5.19 is plotted in Figure 5.7. Note that the vertical axis has been magnified 1000 times.

The electric field distributions of the second lowest mode computed at $\beta/k_0 \approx 1.07$ are displayed in Figure 5.8, where the discontinuity of the E_x component is not plotted properly due to limitations of the plotting program rather than the FEM package. Also, in the same figure, the E_y component, which should be symmetric theoretically, is somewhat asymmetric. This is because E_y is a nondominant component and thus is more vulnerable to numerical errors. As E_y is almost one order in magnitude less than the dominant E_x component, the slight asymmetry of E_y does not have much effect on the overall solution, and this is confirmed by plotting its magnitude distribution (or norm) in Figure 5.9, where no obvious asymmetry can be observed.

Finally, the calculated relative residual (Equation 5.17) of the first two lowest modes is plotted in Figure 5.10, where the vertical axis has also been magnified 1000 times. It should be mentioned that the relative errors and the relative residuals in Figures 5.7 and 5.10, though very small, could be further reduced if the termination of the iterations for the solutions was deferred by using a stricter convergence criterion even with the present single precision computations, although one cannot expect significant improvements in the eigenpairs computed in this way because of the limited number of degrees of freedom employed in the FEM.

In this work the eigenvalue convergence criterion is defined by

$$\frac{\lambda_{i+1} - \lambda_i}{\lambda_{i+1}} \le \varepsilon \tag{5.23}$$

where λ is the eigenvalue, and i is the iteration index. For the above results, $\varepsilon = 1.0 \times 10^{-8}$ was chosen for the fundamental mode (LSE_{10}), and $\varepsilon = 3.0 \times 10^{-7}$ was chosen for the second lowest mode (LSM_{11}).

5.6 Nonlinear Acceleration Techniques

The solution procedure for nonlinear algebraic equations in the present work is iterative in nature and thus can be accelerated. The argument to be accelerated can be either the eigenvalue or the eigenvector but not

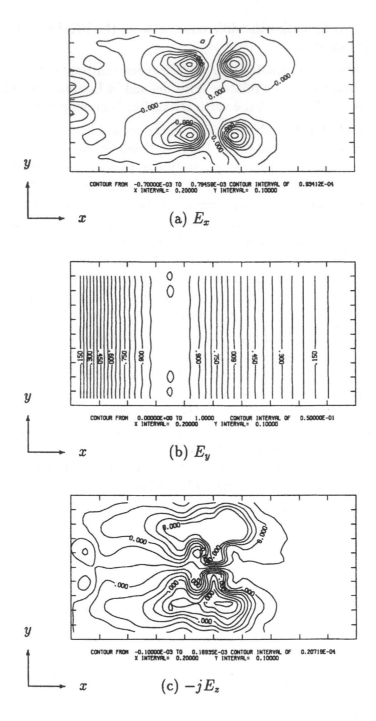

Figure 5.5: Electric field distributions of the fundamental mode (LSE_{10}) at $\beta/k_0 \approx 1.20$.

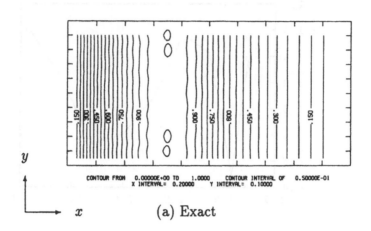

(a) Exact

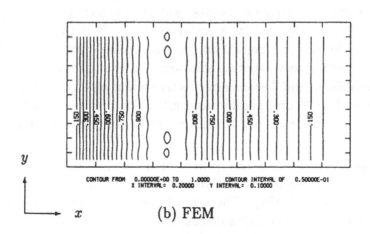

(b) FEM

Figure 5.6: The electric field magnitude distribution of the fundamental mode (LSE_{10}) at $\beta/k_0 \approx 1.20$: (a) the analytical solution; (b) the FEM solution.

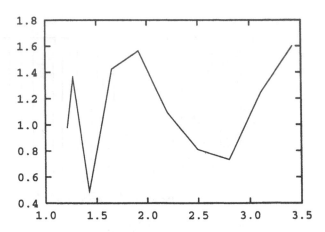

Figure 5.7: The electric field relative error of the fundamental mode of the dielectric-loaded waveguide computed by the FEM: $e_r \times 1000$ *vs* $k_0 a$.

both if one does not want to run the risk of failing to obtain convergence. For the eigenvalue the best known acceleration technique is the *Aitken's δ^2-method* [181] for a scalar quantity. It is the author's experience that this scalar acceleration process is quite effective only when one does not need a very accurate solution, *say* less than 4 significant digits for eigenvalues, and becomes ineffective when eigenvalues are expected to have at least 6 or 7 significant digits. The results quoted in [24] also reflect this. Therefore, our attention will be confined to the acceleration of eigenvectors.

Each component of an eigenvector is a scalar quantity and therefore may be accelerated by the Aitkin's δ^2-method, which leads to the *vector Aitkin's δ^2-method* [181]. There are inherent problems in applying the vector Aitkin's δ^2-method as discussed in [77, 155]. The modified version of the vector Aitkin's δ^2-method described by Zienkiewicz and Irons in [182] has been found not always to give convergence. The Aitkin's δ^2-method is equivalent to an adaptive overrelaxation method, but in this work a fixed overrelaxation factor for the acceleration of the nonlinear iteration procedure is employed rather than an adaptive one, in order to stabilise the nonlinear iteration process.

Let $\boldsymbol{x}_k$ ($k = 1, 2, 3, \ldots$) be an eigenvector obtained at the kth step

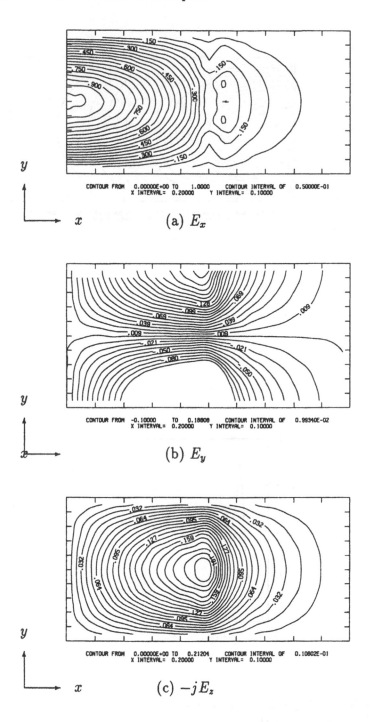

Figure 5.8: Electric field distributions of the second lowest mode (LSM_{11}) at $\beta/k_0 \approx 1.07$.

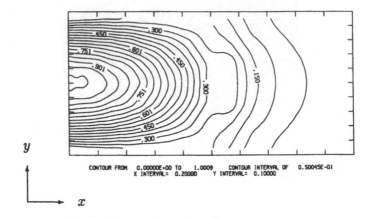

Figure 5.9: The electric field magnitude distribution of the second lowest mode (LSM_{11}) at $\beta/k_0 \approx 1.07$.

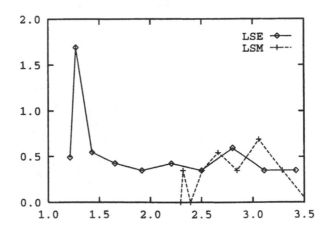

Figure 5.10: The relative residual in the FEM computation of the first two lowest modes of the dielectric-loaded waveguide: $r_r \times 1000$ *vs* $k_0 a$.

of the nonlinear iteration procedure after stage **b)** and assume it is normalised so that

$$\|\boldsymbol{x}_k\|_\infty \; = \; 1.0. \tag{5.24}$$

The eigenvector $\boldsymbol{x}_k$ is denoted by $\bar{\boldsymbol{x}}_k$ after the acceleration and renormalisation. Then the acceleration technique is described by

$$\boldsymbol{y}_k \; = \; \bar{\boldsymbol{x}}_{k-1} + A_{cc}\left(\boldsymbol{x}_k - \bar{\boldsymbol{x}}_{k-1}\right). \tag{5.25}$$

and

$$\bar{\boldsymbol{x}}_k \; = \; \frac{\boldsymbol{y}_k}{\|\boldsymbol{y}_k\|_\infty} \tag{5.26}$$

Here A_{cc} is the nonlinear acceleration factor.

Given the guided power P_g, the eigenvector for the nonlinear problem, $\bar{\bar{\boldsymbol{x}}}_k$, in Equation 3.22 is given by

$$\bar{\bar{\boldsymbol{x}}}_k \; = \; \sqrt{P_g/P(\bar{\boldsymbol{x}}_k)}\; \bar{\boldsymbol{x}}_k \tag{5.27}$$

where $P(\bar{\boldsymbol{x}}_k)$ is the guided power determined by $\bar{\boldsymbol{x}}_k$ according to Equation 2.34.

A fixed acceleration factor is, of course, by no means optimal; however, with a conservative value of around 1.3, the method has proven to be beneficial for all the examples tested during this work. The computation time saved is about 20% typically [183].

Chapter 6

Applications to Nonlinear Optical Waveguides

6.1 Introduction

Nonlinear optical waveguides can behave quite differently from their linear counterparts as the field can manipulate itself via nonlinear interactions [184]-[187]. Care must be exercised in analysing these structures. Although many research papers have been appearing in the literature to target the problem of nonlinear guided waves, several important features and phenomena are either ignored or poorly understood.

To the best knowledge of the author, the rigorous modal analysis of nonlinear optical waveguides in a quasi-3D structure using the full vectorial formulation was first reported in the literature by Hayata-Koshiba [25] who used the H formulation, and by Wang-Binh-Cambrell [188] and Wang-Cambrell-Binh [189] who used the E formulation. In the work of [25] and [188], the Kerr-like nonlinearity model was adopted. Later we discovered that the results in [25] and [188] are valid only when the self-focussing action is weaker than the diffraction effect due to ignorance of a very important aspect [189]. We will confirm this shortly via the example of the classical channel waveguide given in [25] and show how to extend the model to cover strong self-focussing effects, which are vital to nonlinear optical switching devices. Also, please refer to [189] for an example of a nonlinear thin-film structure.

The vector finite element solution of saturable nonlinear strip-loaded optical waveguides was reported by Ettinger, *et al* [153], where the H

95

formulation is employed. Though the nonlinear permittivity model in [153] is valid for strong nonlinear effects, the author was concerned solely with saturable nonlinear effects and, like the authors of other papers [190]-[192] on scalar nonlinear wave equations, did not conclude that the saturable model is also essential mathematically. Moreover, no explanation was given for the "jump" appearing in the power dispersion curve, just as none was mentioned in the early work [25, 188, 189].

To simulate *electrically*-nonlinear optical effects, as discussed in Chapter 2, a disadvantage of the H formulation is that the modified nonlinear permittivity during the iteration process is discontinuous across inter-element interfaces whenever the FEM with C^0-elements is applied, and this discontinuity might cause errors to accumulate during the nonlinear iteration process. With the E formulation, as discussed in Chapter 3, difficulties arise in handling the discontinuity at dielectric interfaces, especially at a corner where the normal component and one of the tangential components cannot be defined. It is of great interest to compare these two formulations where possible so as to check their mutual consistency, which will be performed for a specific example of a nonlinear optical switching structure in § 6.3. To make the power dispersion curve more universal, several normalisation procedures are incorporated.

For some nonlinear optical waveguides exhibiting sharp switching phenomena, there exists a jump in the power dispersion curve, β/k_0 *vs* P, at a certain power as shown in [25, 153, 188, 189]. Here we will propose a non-conventional solution strategy to simulate the jump and show that the jump corresponds to an interesting bistability phenomenon. The propagation stability of the nonlinear modes will be investigated in the next chapter. With a better understanding of the jump, it will be demonstrated how to modify the conventional solution technique to compute all the stable nonlinear modes. A summary of these results has been reported in [193, 194].

Among various nonlinear optical waveguides, planar (interface, thin-film, slab and multilayer) structures have been extensively investigated due to their ease of fabrication and analysis. In analysing these structures, it is customary to assume that both the refractive index and the guided field are uniformly distributed along the dielectric interfaces so that TE and TM modes (described by a scalar function of one spatial variable) can be easily found by either analytical [40]-[43], [195]-[216] or numerical [23, 24], [217]-[220] techniques. It is well known that the quasi-2D formulation of planar structures works fairly well for practical problems in the case of linear waveguides. However, one should bear in mind that a highly intense field may form in a guiding channel made from a self-focussing nonlinear medium. Hence such planar structures should be formulated in quasi-3D to simulate strong nonlinear effects.

Thin-film structures have been formulated in quasi-3D with the weak-guidance approximation (described by a scalar function of two spatial variables) [190]-[192]. In a nonlinear situation, however, these results should be checked by a rigorous formulation. Furthermore, some of the results in [190]-[192] are not reliable for the same reason discussed previously for the classical channel structure. We have already reported the preliminary results of the comparison of scalar and vector formulations of nonlinear thin-film structures in quasi-3D [189]. However, the jump in the power dispersion curve was ignored in that paper. Here it is intended to give a full account of how to correctly simulate nonlinear thin-film structures.

Many reference works continued to appear while this book was being completed, for example, [221]-[231].

6.2   The Necessity of including Saturation in the Nonlinear Permittivity Model

The structure considered here is the classical channel waveguide structure consisting of a rectangular core bounded by a cladding [232], as is shown in Figure 6.1.

We consider the case of a linear core bounded by a self-focussing nonlinear cladding. The nonlinear perturbation term of the relative permittivity dyadic in Equation 2.3 is assumed to be

$$\hat{\epsilon}_r^n = \begin{bmatrix} \epsilon_1^n & 0 & 0 \\ 0 & \epsilon_2^n & 0 \\ 0 & 0 & \epsilon_3^n \end{bmatrix}. \tag{6.1}$$

For Kerr-like nonlinearity,

$$\epsilon_i^n = a(|E_i|^2 + b \sum_{\substack{j=1 \\ (j \neq i)}}^{3} |E_j|^2), \qquad i = 1, 2, 3 \tag{6.2}$$

where a [m^2/V^2] is the nonlinear coefficient defined by

$$a = c_0 \epsilon_0 \epsilon_r^l n_2 \tag{6.3}$$

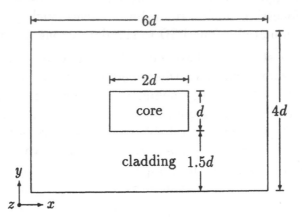

Figure 6.1: The classical structure of a rectangular channel waveguide with electric wall boundaries.

in terms of the usual *nonlinear optical coefficient* n_2 [m^2/W], the velocity of light in vacuum c_0, the vacuum permittivity ϵ_0 and the linear term of the relative permittivity ϵ_r^l which is assumed to be isotropic; and b depends on the origin of the nonlinear mechanism [38]:

$$b = \begin{cases} 1 & \text{electrostriction and heating} \\ \frac{1}{3} & \text{electronic distortion} \\ -\frac{1}{2} & \text{molecular orientation} \end{cases} \qquad (6.4)$$

To facilitate comparison with Ref. [25], the other parameters are chosen as follows: the free-space wavelength $\lambda_0 = 0.515$ μm ($k_0 = 2\pi/\lambda_0$), $n_2 = 1.0 \times 10^{-9}$ m^2/W, $\beta d = 10$ and

$$\epsilon_r^l = \begin{cases} 1.55^2 & \text{core} \\ 1.52^2 & \text{cladding} \end{cases} \qquad (6.5)$$

The structure in Figure 6.1 is meshed into 116 second-order elements with 265 nodes, as shown in Figure 6.2. The maximum node index difference between any two related nodes is 37 after node renumbering. The fundamental mode of the waveguide under investigation corresponds to the H_{11}^x or E_{11}^y mode in the linear case, which is illustrated in Figure 6.3.

As the structure has been normalised with respect to the propagation constant β, the eigenvalue λ in the generalised eigenvalue equation

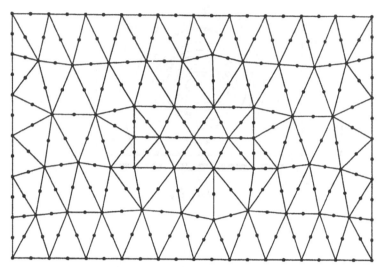

Figure 6.2: Mesh network of the rectangular channel waveguide structure (116 second-order elements, 265 nodes).

now represents $(k_0/\beta)^2$. The power dispersion curve can be found in the conventional way by specifying the guided power P and calculating β/k_0 [25, 183, 188]. The computed power dispersion curve for the isotropic case of $b = 1$ from the present work together with the one obtained by Hayata and Koshiba in Ref. [25] are shown in Figure 6.4[1], both using the H formulation. When the power is below a certain threshold, excellent agreement between the two curves is obtained. When the power is higher, however, a significant discrepancy is observed. We have found that this discrepancy is very sensitive to the element size in the nonlinear region, indicating that the mesh employed is not fine enough to achieve adequate accuracy.

To check the effect of the size of the elements on the solution, the magnetic field distributions at $P = 80$ μW are illustrated in Figure 6.5, where the field converges to quite a small spot involving a very limited number of nodes. Therefore, the results obtained are not reliable and should be justified by the convergence with further refinement of the mesh. Figure 6.6 illustrates our refined mesh, consisting of 176 second-

---

[1] All the powers in the power dispersion relations in Ref. [25] should be scaled down by a factor of 2 due to a programming error [233], and the corrected values are used here to facilitate the comparison.

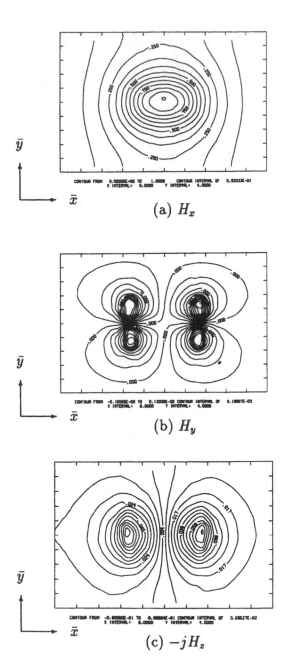

Figure 6.3: The magnetic field distributions of the fundamental mode (H_{11}^x or E_{11}^y) for the rectangular channel waveguide in the linear case.

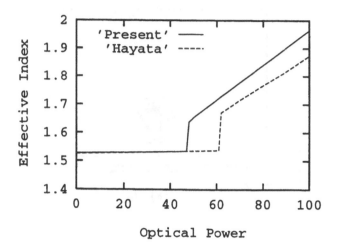

Figure 6.4: The power dispersion curve of the fundamental nonlinear mode: β/k_0 *vs* P [μW].

order elements with 389 nodal points (the maximum node index difference between any two related nodes is 48), for the subsequent computation.

The power dispersion curves corresponding to both the original and the refined meshes are plotted in Figure 6.7. When the power is below the threshold, the two curves are indistinguishable, as expected. When the power is above the threshold, however, significant discrepancies are observed, and the nonlinear iteration procedure fails to give convergence when $P \geq 70$ μW. A typical example of the convergence behaviour (at $P = 80$ μW) with both meshes is shown in Figure 6.8. With the original mesh the convergence behaviour is monotonic, which is similar to that of Ref. [25] except that the convergence in the the present work is about twice as fast; but with the refined mesh, the eigenvalue sought fails to converge with further iterations and the maximum value of the effective index, reached after the 23rd iteration, is chosen as the representative in Figure 6.7. The instability is understandable as the field depends only on such a small portion of the nodes, as shown in Figure 6.9, that slight numerical round-off error will have a great effect on the convergence due to the nonlinear iteration process. We envisage that the authors of Ref. [25] would also have observed the same phenomenon if they had

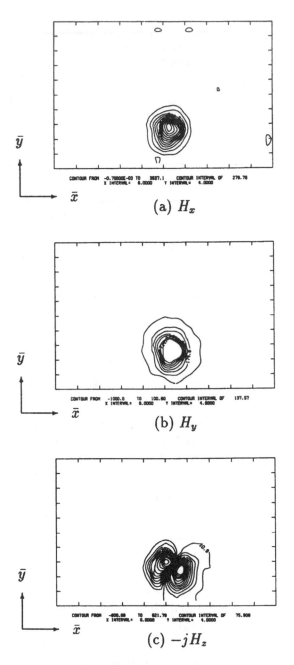

Figure 6.5: The magnetic field distributions of the fundamental nonlinear mode at $P = 80$ μW using the mesh shown in Figure 6.2.

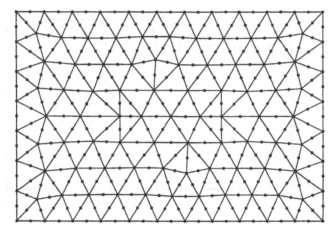

Figure 6.6: The refined mesh network of the rectangular channel wave-guide structure (176 second-order elements, 389 nodes).

performed the computation on a much finer mesh.

Now the discrepancy between the upper branches of the two power dispersion curves in Figure 6.4 is quite understandable since different meshes were used in each work. It remains to answer why the threshold power predicted in the present work is apparently less than that predicted in Ref. [25] as the threshold power is not very sensitive to variations of the mesh employed, as shown in Figure 6.7. We found that the acceleration technique employed in the present work has only a slight effect on the threshold power calculated; but the *convergence criteria* established for completing a linear solution during each step of the nonlinear iteration process and for terminating the nonlinear iteration process have a great bearing. The reason is that the convergence rate is very slow due to the competition between the linear and the nonlinear regions when the guided power is not considerably higher than the threshold, and the nonlinear iteration may be terminated prematurely if a loose convergence criterion is adopted. We believe that a stricter convergence criterion for the nonlinear iteration is used in the present work as compared with the one used in Ref. [25]. Also, the higher convergence rate (in terms of the number of nonlinear iterations) achieved in the present work is most probably caused by a stricter convergence criterion for each linear solution step. In this section of our work the convergence criteria for completing each linear solution step and for terminating the nonlinear

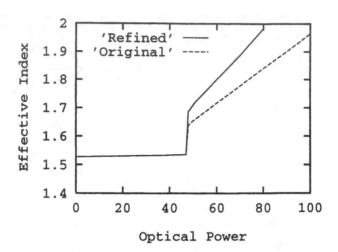

Figure 6.7: The power dispersion curves of the fundamental nonlinear mode corresponding to both the original and the refined meshes: β/k_0 vs P [μW].

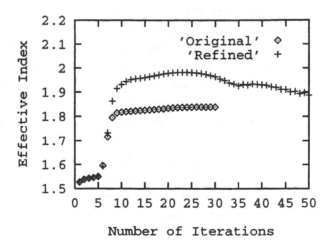

Figure 6.8: Convergence behaviour of the effective index during the non-linear iteration procedure for $P = 80$ μW with both the original and the refined meshes: β/k_0 vs N (number of iterations).

iteration process are the same, namely

$$\varepsilon = 1.0 \times 10^{-6} \tag{6.6}$$

where ε has been defined in Equation 5.23.

As shown in Figure 6.9, the self-focussed magnetic field distribution at $P = 80$ μW with the finer mesh is obviously more confined than the one with the coarser mesh, shown in Figure 6.5, even though the node spacing is only about 13% less. To compare the magnitude distributions of the magnetic field associated with these different meshes, the magnitudes $\|H\|_2$ (*i.e.* $\sqrt{|H_x|^2 + |H_y|^2 + |H_z|^2}$) corresponding to both meshes are plotted in Figure 6.10, where the confinement of the fields is limited by the finite size of the elements and the limited number of degrees of freedom per element. Further refinement of the mesh fails to give convergence. The relative residuals associated with the computation of the power dispersion curves with both meshes are given in Figure 6.11, which have the same order of magnitude as those of the linear example in Chapter 5.

There is no difficulty in explaining the above phenomenon. With increasing power, the self-focussing action will overcome the diffraction action, and the intensely self-focussed field will form a self-guiding channel in the nonlinear region of positive nonlinearity ($n_2 > 0$). The intensely self-focussed field will increase the local permittivity via the nonlinear interaction, and the increased permittivity will further focus and strengthen the local field. This self-focussing process will continue without end for the Kerr-like nonlinearity model. As is known, the Kerr-like nonlinearity is only an approximation for weak nonlinear effects and it is misleading to apply it to any strong nonlinear process.

> *When the field energy density increases without limit in exact focussing, there is no justification for retaining only the lowest (the third) power of the nonlinearity* [39].

Apart from this *physical* requirement of including higher-order nonlinear terms for the solution to make sense, *mathematically* it seems necessary to incorporate those higher-order terms or at least adopt a saturating nonlinear permittivity model in order to achieve numerical convergence, or more precisely, to allow the nonlinear electromagnetic wave equation to possess a realistic solution. Consequently, the results of Refs [25, 188, 191, 192] corresponding to the simple Kerr-like nonlinear permittivity

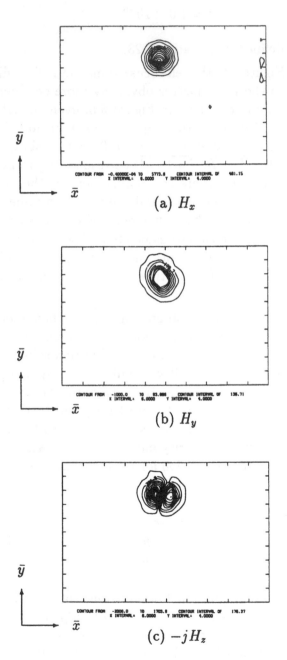

Figure 6.9: The magnetic field distributions of the fundamental nonlinear mode at $P = 80$ μW with the refined mesh after 23 nonlinear iterations.

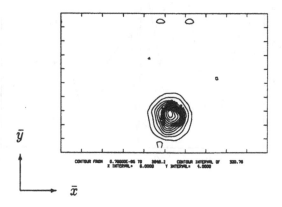

(a) $\|\boldsymbol{H}\|_2$ resulting from the original mesh

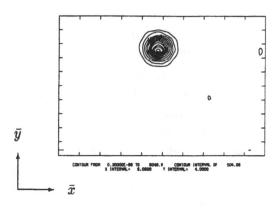

(b) $\|\boldsymbol{H}\|_2$ resulting from the refined mesh after 23 nonlinear iterations

Figure 6.10: A comparison of the magnetic field magnitude distributions of the fundamental nonlinear mode at $P = 80$ μW resulting from both the original and the refined meshes.

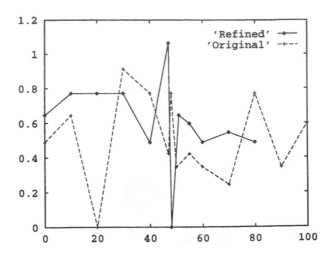

Figure 6.11: The relative residuals in the computation of the fundamental nonlinear mode using both the original and the refined meshes: $r_r \times 1000$ vs P [μW].

model when simulating strong nonlinear effects in quasi-3D structures are improper, both *physically* and *mathematically*.

The self-focussing of the field, to the extent that the nonlinearity is no longer small, has a great bearing on all-optical switching and bistability. Therefore, it is of practical importance to extend the simple Kerr-like model to cover strong nonlinear effects. Instead of incorporating further higher-order nonlinear terms, whose coefficients are usually not available, it is more convenient to adopt a saturating nonlinear permittivity model as all practical materials have limits on the maximum permittivity that may be induced before optical damage.

Nonlinear permittivity models with saturation have been employed by many workers to simulate nonlinear guided wave phenomena, *e.g.* [40]-[43], [183], [189]-[192], [234, 235]. Apart from our work [183, 189, 234], in these publications a saturable permittivity model was employed either to simulate saturating nonlinear effects as such or to obtain the "N-shaped" power dispersion curve; none of them concluded that *mathematically* it is essential to incorporate saturation into the nonlinear permittivity model in a quasi-3D structure in order that the nonlinear wave equation shall possess a realistic solution when the self-focussing action becomes dominant.

Broadly, there are two main saturation models appearing in the literature [40, 41, 43, 153, 213]:

$$\epsilon_r^n = \frac{a|E|^2}{1 + \alpha a|E|^2} \tag{6.7}$$

and

$$\epsilon_r^n = \frac{1}{\alpha}[1 - \exp(-\alpha a|E|^2)] \tag{6.8}$$

where α is the saturation parameter and a is defined in Equation 6.3. Without loss of generality, we will adopt the former, which models the properties of a two-level system far from resonance [40] and is also a good approximation for some semiconductors and other materials [235]. As Equation 6.7 is valid only for scalar fields, we have to extend this model to the vector field case. By analogy with the scalar model, we may write the saturation model for vector fields as

$$\epsilon_i^n = \frac{a f_i(\boldsymbol{E})}{1 + \alpha a f_i(\boldsymbol{E})} \qquad i = 1, 2, 3 \tag{6.9}$$

and use this model to replace the Kerr-like model in Equation 6.2, where "$a f_i(\cdot)$" is assumed to be the same as ϵ_i^n in Equation 6.2. For the isotropic case of $b = 1$ in Equation 6.2, the saturation model becomes

$$\epsilon_i^n = \frac{a\|E\|_2^2}{1 + \alpha a\|E\|_2^2} \qquad i = 1, 2, 3 \tag{6.10}$$

which will be used in subsequent computations.

Now we consider the numerical convergence behaviour with respect to the refinement of the mesh by using the saturation model with $\alpha = 3$ without loss of generality though practical values of the saturation parameter are usually much higher than this value [3], [236]-[238]. The power dispersion curves associated with both the original and the refined meshes are shown in Figure 6.12. Apparently convergence is readily achieved with a saturating model. The slight discrepancy in the threshold power indicates that the original mesh (and possibly the refined one as well) is not yet fine enough to determine the threshold accurately. The associated relative residuals in computing the dispersion relations are plotted in Figure 6.13, which have the same order of magnitude as those of the linear example in Chapter 5. Therefore, the plotting of residuals will be omitted for brevity in the rest of the chapter.

The magnetic field distributions at power levels below the threshold are indistinguishable from those without saturation, as expected. When the guided power exceeds the threshold, however, the field distributions are surprisingly different from those without saturation. Typical examples of the magnetic field at $P = 80 \ \mu$W are shown in Figures 6.14 and 6.15, corresponding to the original and the refined meshes, respectively, between which little difference can be observed except that the intense self-focussing may occur in either the upper or the lower region of the cladding, depending on the numerical round-off, due to the geometrical symmetry of the structure. The field patterns in Figure 6.14 and 6.15 are surprising but reasonable:

- The saturation effect makes the field pattern more diffuse;

- The intense self-focussing action enables the guided wave to be self-channelling and therefore independent of the core region;

- The order of this "half-pattern" mode, which makes use of the electric wall as a symmetry plane, is lower than that of the corresponding "full-pattern" mode, if it exists, for the given closed structure with fixed power; therefore the mode obtained here is consistent with the solution method for finding the lowest mode.

Finally, one might have noticed that the symmetric structure can support asymmetric modes when strong self-focussing occurs. The reason is quite simple: in nonlinear materials, symmetry at low guided powers does not mean symmetry at very high powers as the field itself can alter the distribution of the permittivity in nonlinear media. Consequently, the so-called "even (or symmetric)" and "odd (or antisymmetric)" modes in [24, 191] most probably no longer exist physically when the guided power exceeds a certain critical value.

To summarise this section, we conclude:

- In simulating strong nonlinear effects in quasi-3D structures, it is essential to incorporate saturation into the nonlinear permittivity model (or at least to include certain higher-order nonlinear terms in the formulation) for both physical and mathematical reasons.

- Symmetric structures can support asymmetric modes in a nonlinear situation. Thus, it is not recommended that use be made of the symmetry of a structure unless the field pattern being sought is known to be symmetric *a priori*.

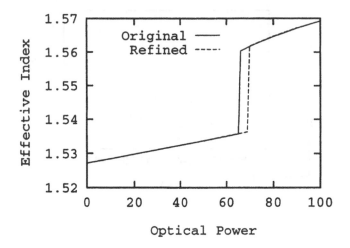

Figure 6.12: The power dispersion curves of the fundamental nonlinear mode corresponding to both the original and the refined meshes for the model with saturation: β/k_0 *vs* P [μW].

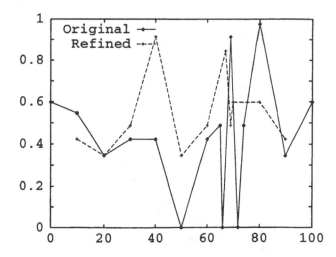

Figure 6.13: The relative residuals in the computation of the fundamental nonlinear mode using both the original and the refined meshes for the saturating permittivity model: $r_r \times 1000$ *vs* P [μW].

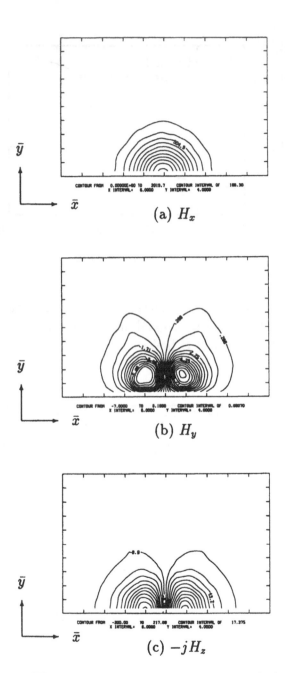

Figure 6.14: The magnetic field distributions of the fundamental non-linear mode for the saturating model at $P = 80$ μW with the original mesh.

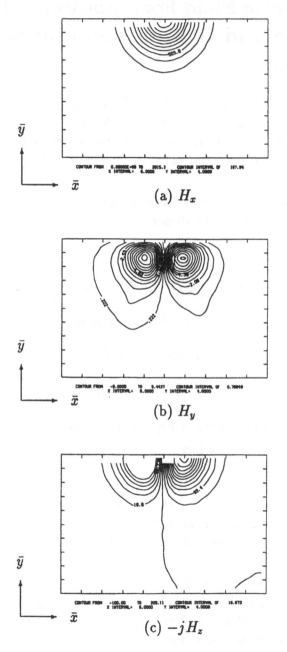

(a) H_x

(b) H_y

(c) $-jH_z$

Figure 6.15: The magnetic field distributions of the fundamental non-linear mode for the saturating model at $P = 80 \ \mu$W with the refined mesh.

6.3 Comparison of the Electric and the Magnetic Field Formulations, Normalisation and Other Computational Aspects

In the previous section, a specific value of the nonlinear optical coefficient was used to facilitate comparison with Ref. [25]. Here we introduce certain normalisation procedures to make the power dispersion curve independent of the nonlinear optical coefficient and therefore more universal.

For the saturating nonlinear isotropic permittivity model given in Equation 6.10, we introduce the dimensionless field variables defined by

$$\widetilde{E} \triangleq \sqrt{a}E = \frac{1}{\sqrt{Z_0}}\sqrt{\epsilon_r^l n_2}E \tag{6.11}$$

$$\widetilde{H} \triangleq Z_0\sqrt{a}H = \sqrt{Z_0}\sqrt{\epsilon_r^l n_2}H \tag{6.12}$$

Consequently, the relation between $\widetilde{E}$ and $\widetilde{H}$ is governed by

$$\widetilde{H} = j\frac{\rho}{k_0}\hat{\mu}_r^{-1} \cdot \overline{\nabla} \times \widetilde{E} \tag{6.13}$$

$$\widetilde{E} = -j\frac{\rho}{k_0}\hat{\epsilon}_r^{-1} \cdot \overline{\nabla} \times \widetilde{H} \tag{6.14}$$

and the saturation model given in Equation 6.10 now becomes

$$\epsilon_i^n = \frac{\|\widetilde{E}\|_2^2}{1 + \alpha\|\widetilde{E}\|_2^2} \qquad i = 1, 2, 3 \tag{6.15}$$

which is explicitly independent of the nonlinear coefficient a defined in Equation 6.3. When the nonlinear medium is inhomogeneous, the largest value of n_2 is used for the normalisation. If the medium is entirely linear ($n_2 = 0$), then $\sqrt{a}$ should be replaced by a constant (of unit value, for example) having dimensions of m/V.

With the field variables normalised, the corresponding guided power P given in Equation 2.34 should be normalised as well. The normalised guided power $\widetilde{P}$, which is also dimensionless, is defined by

$$\begin{aligned}
\widetilde{P} &\triangleq 2Z_0 a k_0^2 P \\
&= \frac{k_0}{\rho} \Im\{-\int_\Omega \widetilde{V}^* \times (\hat{p}^{-1} \ (\overline{\nabla} \times \widetilde{V})) \cdot a_z \ d\tilde{x}d\tilde{y}\}
\end{aligned} \tag{6.16}$$

where $\widetilde{V}$ denotes either $\widetilde{E}$ or $\widetilde{H}$.

The formulations described in Chapter 2 are still valid for the normalised field variable $\widetilde{V}$ except that the nonlinear term of the relative permittivity is expressed in terms of $\widetilde{E}$; and the computation proceeds in the same way as before except that only α needs to be specified for nonlinear problems without referring to any particular value of the nonlinear coefficient a provided that the normalised guided power is also used.

So far it has been assumed that only one homogeneous nonlinear medium is involved having a constant nonlinear coefficient a. In situations when multiple nonlinear media are involved or the nonlinear coefficient a is a function of position, the field variables will be normalised with respect to the largest value of a in the whole structure. Consequently, merely the *relative* value of any nonlinear coefficient will be used in the computation, *i.e.*, the actual nonlinear coefficient divided by the largest nonlinear coefficient.

The objectives of the comparison between the E and the H formulations are threefold. Firstly, it is essential to check the mutual consistency of the results from both formulations as each formulation has its own inherent difficulties, as discussed previously. Secondly, it is of practical importance to compare the efficiency of the two formulations as computations of nonlinear structures are rather expensive. And finally, it would be of great value to know how well the E formulation behaves in a nonlinear situation as it is noted for its inaccuracy (due to the discontinuity of the normal component at dielectric interfaces) in a linear situation.

In order to compare results from the present work with those of Ref. [25], the structure in the previous section was normalised with respect to β. Here we adopt the alternative approach of normalising the structure with respect to k_0. There is a good reason for structures to be normalised in this way, because, in practice, propagation properties are to be analysed with the operating frequency fixed.

Figure 6.16 shows the geometry of a nonlinear film-loaded ion-exchanged-channel waveguide, consisting of an ion-exchanged channel in a glass substrate and a nonlinear overlay [183]. The relative permittivity

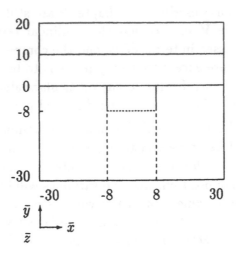

Figure 6.16: A nonlinear film-loaded ion-exchanged-channel waveguide structure and coordinate system.

dyadic is taken to be

$$
\hat{\epsilon}_r = \begin{cases}
\hat{I} & |\bar{x}| \le 30, \quad 10 < \bar{y} \le 20 \\
\left(1.53^2 + \dfrac{\|\widetilde{E}\|_2^2}{1 + \alpha\|\widetilde{E}\|_2^2}\right)\hat{I} & |\bar{x}| \le 30, \quad 0 < \bar{y} \le 10 \\
(1.54 + 0.045\, erfc(-\bar{y}/8))^2\,\hat{I} & |\bar{x}| \le 8, \quad -30 \le \bar{y} \le 0 \\
1.54^2\hat{I} & 8 < |\bar{x}| \le 30, \quad -30 \le \bar{y} \le 0
\end{cases}
$$

(6.17)

where $\hat{I}$ is the identity dyadic, and the boundary conditions are assumed to represent perfect conductors. The saturation parameter α is taken to be 1.0 without loss of generality.

The whole structure is meshed into 126 second-order elements with 281 nodal points, as shown in Figure 6.17. After node renumbering, the maximum node index difference between any two related nodes is 43.

The power dispersion relations of the fundamental mode resulting from both the **E** and the **H** formulations are plotted in Figure 6.18. When $\widetilde{P} < 50$, the results from both formulations are in good agreement; therefore their mutual consistency is verified. When $\widetilde{P} > 50$, a slight discrepancy is observed and the discrepancy increases as the guided power increases. With increasing guided power, the field distribution becomes more confined and a finer mesh is required to achieve a certain accuracy.

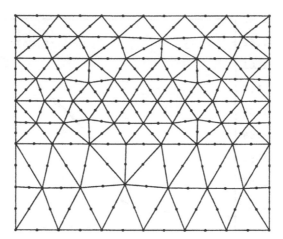

Figure 6.17: Mesh network of the film-loaded ion-exchanged-channel waveguide structure (126 second-order elements, 281 nodes).

Here the discrepancy simply indicates that there is some uncertainty in the results obtained probably due to the rather small saturation parameter adopted or the rather coarse mesh employed.

Apart from accuracy, the convergence behaviour and the computation time are also important since the computation time may be traded for the accuracy. The number of nonlinear iterations (NNI) required for computations of the fundamental mode for each formulation is plotted in Figure 6.19. On average, the H formulation gives faster convergence. In this section the convergence criteria for both the E formulation and the H formulation are identical. More specifically,

$$\varepsilon = 5.0 \times 10^{-7} \tag{6.18}$$

for completing each linear solution step, and

$$\varepsilon = 1.0 \times 10^{-6} \tag{6.19}$$

for terminating the nonlinear iteration process.

The number of nonlinear iterations required for solutions near the inflection point of the power dispersion curves is striking. For algorithms whose computation time is proportional to the number of nonlinear iterations, the total computation time required is enormous and probably

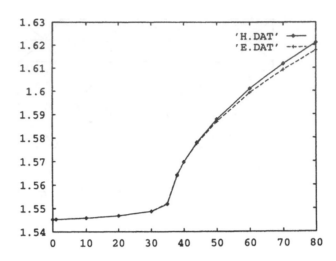

Figure 6.18: A comparison of the power dispersion curves resulting from the electric and the magnetic field formulations: β/k_0 vs $\tilde{P}$.

a supercomputer is needed. The computation in the present work, however, is performed by the SOR-RQ technique to solve the linear generalised eigenvalue problem, in which the most recent information is fully utilised at each step of the nonlinear iteration procedure. Consequently, the computation effort is very reasonable. The CPU time[2] used on a VAX8700 mainframe computer is shown in Figure 6.20. Obviously, the H formulation is preferred.

As the structure has been normalised with respect to k_0, another iteration process (§ 5.2) is incorporated along with the nonlinear iteration procedure. To demonstrate the convergence of our algorithm, all of the above computations were started from rather arbitrary initial conditions for the first nonlinear iteration and convergence was always achieved. We have also tried using different initial conditions and the final results are independent of the initial condition chosen. However, a better initial condition for iteration can lead to less CPU time being used. Therefore it is highly recommended that, for a solution at a given power $\tilde{P}$, the field and the effective index corresponding to the power closest to $\tilde{P}$,

---

[2]This has only a qualitative meaning since the computation is performed on a time-shared system. We have observed that the CPU time used can vary by up to 10% for the same problem being computed.

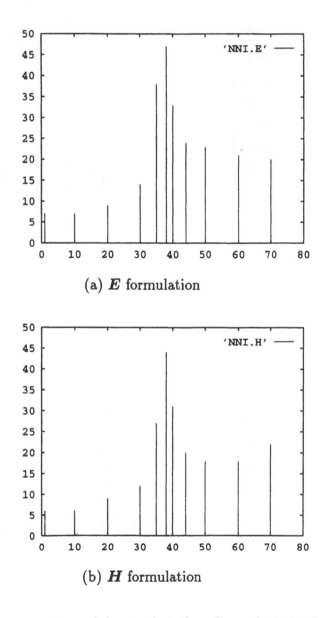

(a) $\boldsymbol{E}$ formulation

(b) $\boldsymbol{H}$ formulation

Figure 6.19: A comparison of the number of nonlinear iterations required for computations of the fundamental mode for each formulation: NNI *vs* $\widetilde{P}$.

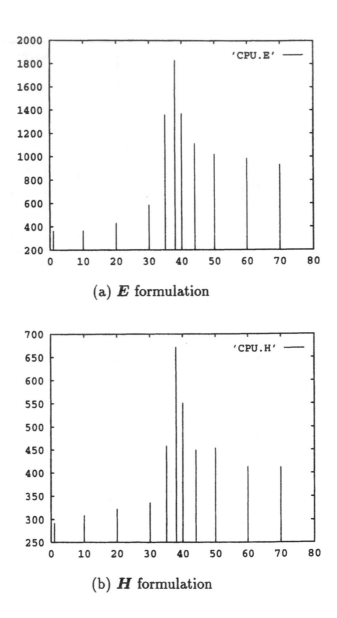

(a) *E* formulation

(b) *H* formulation

Figure 6.20: A comparison of the CPU time used on a VAX8700 computer for computations of the fundamental mode for each formulation: CPU time [seconds] *vs* $\widetilde{P}$.

if available, be used as the initial condition for the nonlinear iteration procedure. For example, if the corresponding linear mode is chosen as the initial condition, the CPU time will be reduced from about (432; 322) seconds to about (357; 90) seconds at $\widetilde{P} = 20.0$ and from about (917; 392) seconds to about (750; 246) seconds at $\widetilde{P} = 80.0$ for the (E; H) formulation. Clearly, for this initial condition, the H formulation is faster than the E formulation. Of course, a better initial condition can further improve the efficiency.

For all the above computations the nonlinear acceleration parameter A_{cc} was taken to be 1.3. It would be interesting to look at the effectiveness of the nonlinear acceleration technique employed. The convergence behaviour with different values of A_{cc} at $\widetilde{P} = 80.0$ is shown in Figure 6.21. To scrutinise the delicate effects on the top part of Figure 6.21, a magnified version is given in Figure 6.22. We found that the convergence rate is not too sensitive within the range of $1.3 \leq A_{cc} \leq 1.5$. The actual CPU time saved by using the nonlinear acceleration technique depends on the convergence criterion established. Generally speaking, the larger the number of effective digits required, the less the saving in CPU time.

For comparison, the field distributions from both formulations for $\widetilde{P} = 20.0$ and 80.0 are illustrated in Figures 6.23-6.26. The non-dominant electric field component E_z at $\widetilde{P} = 20.0$, as can be seen, is somewhat distorted, suggesting that the E formulation is not as accurate as the H formulation for the present example.

At low power ($\widetilde{P} = 20.0$), the field is mainly distributed in the linear region, whereas at high power ($\widetilde{P} = 80.0$), the field is well confined in the nonlinear region due to intense self-focussing, as expected. This phenomenon indicates that the structure analysed may be employed as an all-optical switching device.

In conclusion, a normalisation procedure has been proposed where both the field and the guided power are made dimensionless and independent of the nonlinear coefficient. Therefore, the power dispersion relations obtained are more universal. Also, it has been shown that the results from both the E and the H formulations are mutually consistent, so one can apply either formulation with confidence. As far as the CPU time is concerned to achieve a certain accuracy, the H formulation is preferred. The nonlinear acceleration technique employed is shown to be rather effective. The structure analysed may be used for all-optical switching purposes.

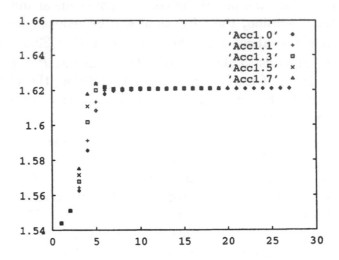

Figure 6.21: The convergence behaviour with nonlinear acceleration $(A_{cc} = 1.1, 1.3, 1.5, 1.7)$ and without acceleration $(A_{cc} = 1.0)$: β/k_0 *vs* NNI.

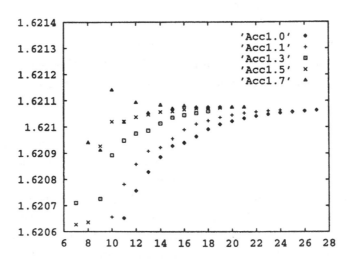

Figure 6.22: A close look at the effectiveness of the nonlinear acceleration technique for $A_{cc} = 1.0, 1.1, 1.3, 1.5, 1.7$: β/k_0 *vs* NNI.

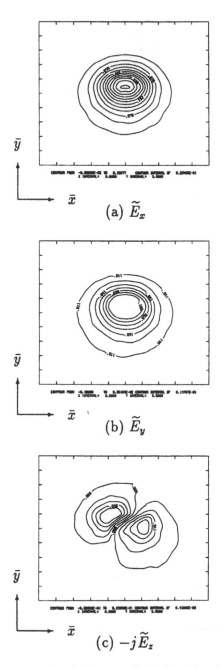

Figure 6.23: The normalised electric field distributions of the fundamental mode at $\widetilde{P} = 20.0$.

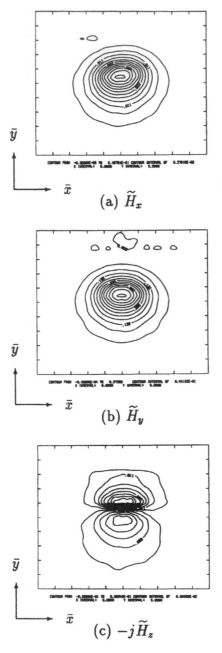

(a) $\widetilde{H}_x$

(b) $\widetilde{H}_y$

(c) $-j\widetilde{H}_z$

Figure 6.24: The normalised magnetic field distributions of the fundamental mode at $\widetilde{P} = 20.0$.

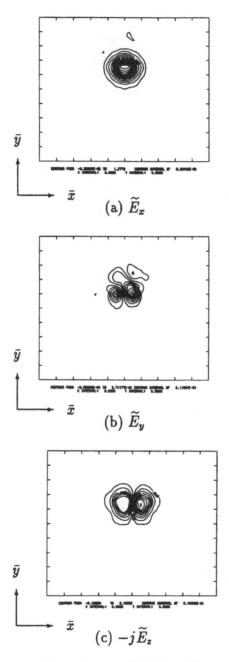

Figure 6.25: The normalised electric field distributions of the fundamental mode at $\widetilde{P} = 80.0$.

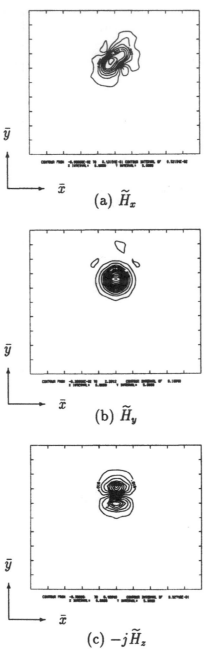

Figure 6.26: The normalised magnetic field distributions of the fundamental mode at $\widetilde{P} = 80.0$.

6.4 Novel Simulation of Bistability Phenomena in Nonlinear Optical Waveguides

In § 6.2 the power dispersion curve exhibits a jump, whereas in § 6.3 the power dispersion curve does not possess such a jump. Comparing these structures and field distributions, we contend that for a jump to exist the dimensions of the nonlinear region should be large enough so that, with strong nonlinear effects, the nonlinear region itself can support a mode. Also the maximum refractive index that the nonlinear medium can reach should be higher than that of the linear guided region so that there exists a sharp switching behaviour rather than a gradual change with increasing power.

When the power is above the threshold, the field distributions given in § 6.2 for the saturating nonlinear permittivity model do not represent any useful physical solution as they confine themselves to the artificial boundary. Prior to investigating the jump phenomenon, we propose a new waveguiding structure shown in Figure 6.27, consisting of a linear guiding channel and a nonlinear loading strip, in which the fundamental mode is expected to be well confined and can never take the boundary as its symmetry plane [194]. The relative permittivity dyadic is taken to be

$$
\hat{\epsilon}_r = \begin{cases}
\hat{I} & |\bar{x}| \leq 30, \quad 15 < \bar{y} \leq 30 \\
 & 8 < |\bar{x}| \leq 30, \quad 0 < \bar{y} \leq 15 \\
(1.52^2 + \dfrac{\|\widetilde{E}\|_2^2}{1 + \alpha\|\widetilde{E}\|_2^2})\hat{I} & |\bar{x}| \leq 8, \quad 0 < \bar{y} \leq 15 \\
1.55^2\hat{I} & |\bar{x}| \leq 8, \quad -12 < \bar{y} \leq 0 \\
1.52^2\hat{I} & 8 < |\bar{x}| \leq 30, \quad -12 < \bar{y} \leq 0 \\
 & |\bar{x}| \leq 30, \quad -30 \leq \bar{y} \leq -12
\end{cases}
\tag{6.20}
$$

Again, $\hat{I}$ is the identity dyadic, and the boundary conditions are assumed to represent perfect conductors, and the saturation parameter α is taken to be 1.0 without loss of generality.

The whole structure is meshed into 126 second-order elements with 281 nodal points, as shown in Figure 6.28. After node renumbering, the maximum node index difference between any two related nodes is 49.

In solving the nonlinear wave equation with finite element methods, it is customary to employ the conventional nonlinear iterative solution

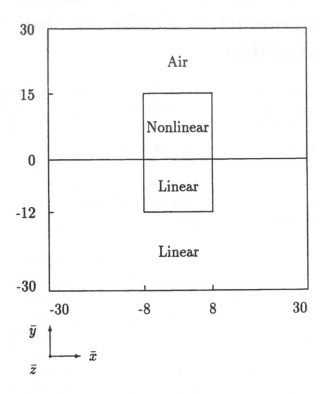

Figure 6.27: The nonlinear strip-loaded channel optical waveguide structure and coordinate system.

technique [25, 153, 188], which can be summarised as follows:

- Seek $(\bar{\beta}; \widetilde{V})$ as the eigenpair for given $\widetilde{P}$.

- Iterate $\bar{\beta}$ and $\hat{\epsilon}_r$ starting from a linear solution.

- Stop when a convergence criterion is satisfied.

The power dispersion curves obtained by the conventional solution technique for both the E and H formulations are shown in Figure 6.29, where, as expected, the curves obtained do possess a jump. This jump phenomenon suggests that the effective index of the mode investigated may be a multivalued function of the guided power. To confirm this prediction, the guided power has to be extracted as the eigenvalue for a specified effective index. To do so, a *non-conventional* nonlinear iteration procedure is proposed as follows.

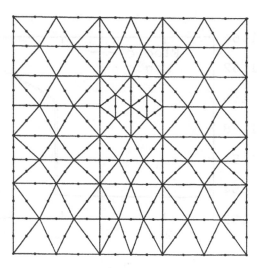

Figure 6.28: Mesh network of the nonlinear strip-loaded channel optical waveguide structure (126 second-order elements, 281 nodes).

- Seek $(\widetilde{P}; \widetilde{V})$ as the eigenpair for given $\bar{\beta}$.

- Iterate $\hat{\epsilon}_r$ starting from either a linear or a nonlinear solution.

- Stop when a convergence criterion is satisfied.

Referring to Equation 6.16, we now introduce a new vector field variable V' and a real scalar coefficient A defined by

$$\widetilde{V} \triangleq AV' \tag{6.21}$$

subject to

$$\frac{k_0}{\rho} \, \Im\{-\int_\Omega V'^* \times (\hat{p}^{-1} \cdot (\overline{\nabla} \times V')) \cdot a_z \, d\bar{x}d\bar{y}\} \equiv 1 \tag{6.22}$$

so that

$$\widetilde{P} \equiv A^2 \tag{6.23}$$

Then $(A^2; V')$ may be solved as the eigenpair by specifying the effective index β/k_0.

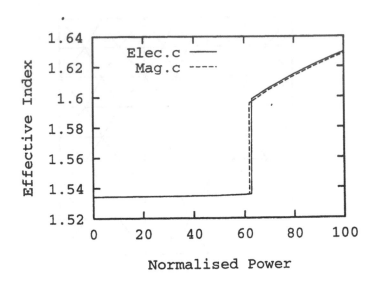

Figure 6.29: Power dispersion curve of the nonlinear strip-loaded optical channel waveguide resulting from the conventional solution technique for both the electric- (solid curve) and the magnetic- (dashed curve) field formulations: β/k_0 vs $\widetilde{P}$.

For simplicity of description, the surface terms resulting from the extended operator formalism will be ignored for the time being. The partial-differential equation corresponding to the final penalty formulation given in Equation 2.29 can be shown to be

$$\overline{\nabla} \times (\hat{p}^{-1} \cdot (\overline{\nabla} \times \boldsymbol{V})) - p\,\hat{q} \cdot \overline{\nabla}\,\overline{\nabla} \cdot (\hat{q} \cdot \boldsymbol{V}) = \lambda\,\hat{q} \cdot \boldsymbol{V} \qquad (6.24)$$

where $\lambda = (k_0/\rho)^2$. This equation is also valid for $\widetilde{V}$ and V'. Note that $\hat{p}$ stands for $\hat{\mu}_r$ in the $\boldsymbol{E}$ formulation or for $\hat{\epsilon}_r$ in the $\boldsymbol{H}$ formulation, while p is the penalty parameter.

Now we consider only the case of the structure being normalised with respect to k_0 as there is no computational advantage in normalising the structure with respect to β in the present situation. Consequently, $\lambda \equiv 1$.

Next, the relative permittivity dyadic is split into a linear term $\hat{\epsilon}_r^l$ plus a nonlinear perturbation term $\hat{\epsilon}_r^n$:

$$\hat{\epsilon}_r \equiv \hat{\epsilon}_r^l + \hat{\epsilon}_r^n. \qquad (6.25)$$

Then the new nonlinear iteration procedure for the solution of Equation 6.24 is given by

$$\overline{\nabla} \times (\hat{\mu}_r^{-1} \cdot (\overline{\nabla} \times \boldsymbol{E}'_{(i+1)})) - p\,\hat{\epsilon}_{r(i)} \cdot \overline{\nabla}\,\overline{\nabla} \cdot (\hat{\epsilon}_{r(i)} \cdot \boldsymbol{E}'_{(i+1)}) - \hat{\epsilon}_r^l \cdot \boldsymbol{E}'_{(i+1)}$$

$$= A_{(i+1)}^2 \{ \frac{\hat{\epsilon}_{r(i)}^n}{A_{(i)}^2} \} \cdot \boldsymbol{E}'_{(i+1)} \tag{6.26}$$

for the $\boldsymbol{E}$ formulation, and by

$$\overline{\nabla} \times ((\hat{\epsilon}_r^l)^{-1} \cdot (\overline{\nabla} \times \boldsymbol{H}'_{(i+1)})) - p\,\hat{\mu}_r \cdot \overline{\nabla}\,\overline{\nabla} \cdot (\hat{\mu}_r \cdot \boldsymbol{H}'_{(i+1)}) - \hat{\mu}_r \cdot \boldsymbol{H}'_{(i+1)}$$

$$= A_{(i+1)}^2 \overline{\nabla} \times \left\{ \left(\frac{(\hat{\epsilon}_r^l)^{-1} - \hat{\epsilon}_{r(i)}^{-1}}{A_{(i)}^2} \right) \cdot (\overline{\nabla} \times \boldsymbol{H}'_{(i+1)}) \right\} \tag{6.27}$$

for the $\boldsymbol{H}$ formulation, subject to Equation 6.22 at each step, where i is the iteration index, and

$$\hat{\epsilon}_{r(i)} \equiv \hat{\epsilon}_r^l + \hat{\epsilon}_{r(i)}^n, \qquad \hat{\epsilon}_{r(i)}^n \equiv \hat{\epsilon}_r^n(A_{(i)}\boldsymbol{E}'_{(i)}), \tag{6.28}$$

and, for the $\boldsymbol{H}$ formulation,

$$\boldsymbol{E}'_{(i)} = \begin{cases} -j\hat{\epsilon}_{r(i)}^{-1} \cdot \overline{\nabla} \times \boldsymbol{H}'_{(i)}, & i = 0 \\ -j\hat{\epsilon}_{r(i-1)}^{-1} \cdot \overline{\nabla} \times \boldsymbol{H}'_{(i)}, & i \geq 1. \end{cases} \tag{6.29}$$

Note that the iteration index $i - 1$ is used to compute $\hat{\epsilon}_r^{-1}$ in Equation 6.29 instead of i when $i \geq 1$. This choice saves a further stage of iteration in evaluating $\boldsymbol{E}'_{(i)}$ from $\boldsymbol{H}'_{(i)}$ at each step of the nonlinear iteration procedure and the resulting algorithm works well. In fact, the above strategy has been employed in the conventional solution technique for the $\boldsymbol{H}$ formulation.

Application of the Galerkin FEM leads to a generalised matrix eigenvalue problem similar to Equation 3.22 except that the term involving the matrix T for the $\boldsymbol{E}$ formulation or the matrix S for the $\boldsymbol{H}$ formulation is split into a linear and a nonlinear part. Here the lowest eigenvalue corresponds to the fundamental mode.

Mathematically, one can seek A as the eigenvalue instead of its square by slightly modifying the right-hand side of Equation 6.26. Note, however, that A^2 has the direct physical meaning of being the normalised guided power.

For the example considered the power dispersion curves computed with the new iteration algorithm for both the $\boldsymbol{E}$ and the $\boldsymbol{H}$ formulations are given in Figure 6.30 together with those presented in Figure 6.29 for comparison. Now it is evident that there is an important bistability phenomenon behind the jump. The stability of the new nonlinear modes will be investigated in the next chapter. This phenomenon can be easily explained:

- When the guided power is relatively low, the refractive index induced by the nonlinearity is not high enough for the nonlinear region to support a mode, and therefore the stable modal field must be mainly distributed in the linear region.

- When the guided power becomes relatively high, the modal field penetrating into the nonlinear region will guide the total field to the nonlinear region by self-focussing. Therefore, the fundamental mode can no longer be confined in the linear region but mainly propagates in the nonlinear region.

- Between these extreme cases, there exists a range of guided powers within which the modal field of the fundamental mode can propagate either mainly in the linear region or mainly in the nonlinear region, depending on the initial condition. In either case the modal field of the fundamental mode is expected to be stable in the sense that the field pattern is maintained as it propagates in the longitudinal direction.

  The intermediate portion (middle branch) of the power dispersion curve of negative slope is expected to be unstable. It serves as a transient "bridge" joining the two stable branches.

Note that the above jump phenomenon was taken for granted in the relevant previous publications [25, 153, 188, 189] except those we reported in [193, 194].

As shown in Figure 6.30, the slope of the lower branch of the power dispersion curve is so small that the accurate computation of that branch with the new algorithm seems *prima facie* impossible. The excellent agreement for the results from the two different solution algorithms on that branch is indeed unexpected.

Typical modal electric fields are given in Figures 6.31-6.33, corresponding to $\tilde{P} \approx 48$ at the three branches. Though they have roughly the same guided power, the field patterns are quite different. Figure 6.31 shows the modal field when it is mainly confined in the linear region;

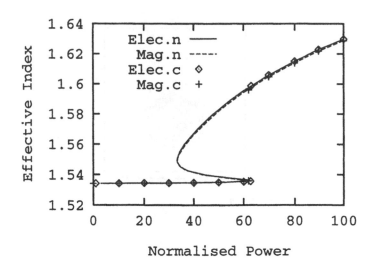

Figure 6.30: Comparison of power dispersion curves (β/k_0 *vs* $\widetilde{P}$) resulting from the conventional and the non-conventional solution algorithms for both the electric- and the magnetic-field formulations, where Elec and Mag denote the electric- and magnetic-field formulations and the extensions c and n denote the conventional and the non-conventional solution algorithm, respectively.

and Figure 6.33 shows the modal field self-focussed in the nonlinear region. Clearly, in Figure 6.32, the modal field corresponding to the middle branch of the power dispersion curve is shared between the linear and the nonlinear regions and shows a tendency towards the latter.

Typical modal magnetic fields are given in Figures 6.34-6.36, corresponding to $\widetilde{P} \approx 48$ at the three branches. Note that the dominant component of the modal electric field is polarised in the x-direction, whereas that of the modal magnetic field is polarised in the y-direction. For this waveguide structure it has been found that the two lowest modes are almost degenerate in the linear case. These results are obtained by choosing favourable initial conditions for the iteration to avoid computational difficulties. Specifically, the E_x and H_y components were set to a non-zero constant and all other field components were set to zero for the first iteration, while the appropriate initial boundary condition was handled by the program itself.

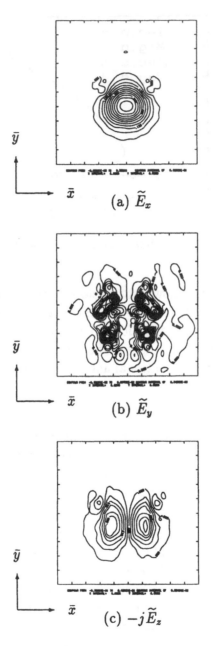

Figure 6.31: The normalised electric field distributions of the fundamental mode on the lower branch of the power dispersion curve for $\widetilde{P} = 48$.

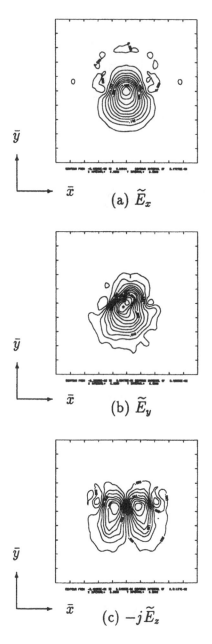

Figure 6.32: The normalised electric field distributions of the fundamental mode on the middle branch of the power dispersion curve for $\widetilde{P} \approx 48$.

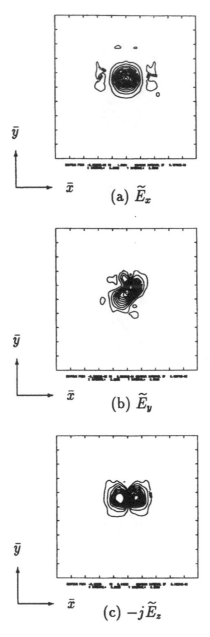

Figure 6.33: The normalised electric field distributions of the fundamental mode on the upper branch of the power dispersion curve for $\widetilde{P} = 48$.

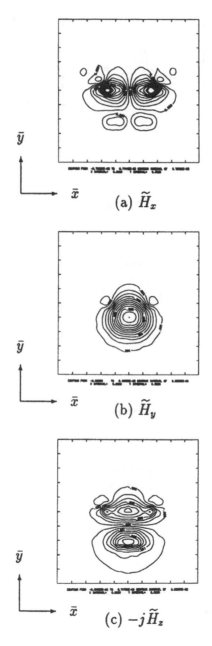

Figure 6.34: The normalised magnetic field distributions of the fundamental mode on the lower branch of the power dispersion curve for $\widetilde{P} = 48$.

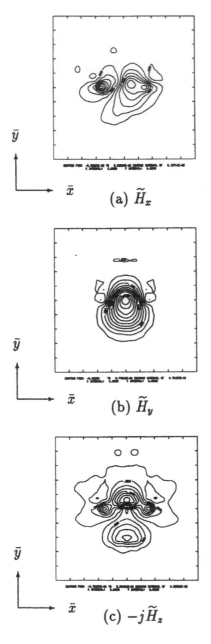

Figure 6.35: The normalised magnetic field distributions of the fundamental mode on the middle branch of the power dispersion curve for $\widetilde{P} \approx 48$.

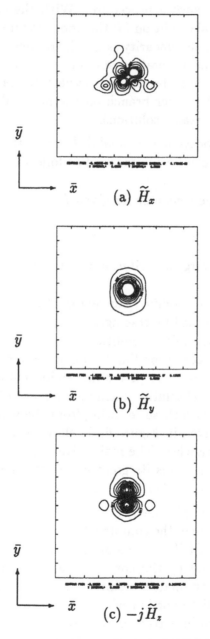

Figure 6.36: The normalised magnetic field distributions of the fundamental mode on the upper branch of the power dispersion curve for $\widetilde{P} = 48$.

We cannot compute any unstable branch by seeking the effective index for given guided power. However, we should be able to find all the stable solutions by such a procedure. With the conventional solution technique the initial condition for the nonlinear iteration is virtually a linear solution as the nonlinearity is not introduced for the first iteration. With the nonlinear iteration starting with a nonlinear solution corresponding to the upper branch, we expect to be able to find the "missing" portion of the upper branch on the power dispersion curve if it really corresponds to stable solutions.

To verify our expectation, a *modified conventional* solution technique is proposed and can be summarised as follows:

- Seek $(\bar{\beta}; \widetilde{V})$ as the eigenpair for given $\widetilde{P}$.

- Iterate $\bar{\beta}$ and $\hat{\epsilon}_r$ starting from a *nonlinear* solution.

- Stop when a convergence criterion is satisfied.

Equipped with the modified conventional solution technique, the computation was performed by tracing up (using a solution corresponding to lower power as the initial condition) the lower branch and tracing down (using a solution corresponding to higher power as the initial condition) the upper branch. The power dispersion curves resulting from the modified conventional solution technique for both the E and the H formulations together with those resulting from the non-conventional solution technique are given in Figure 6.37, showing excellent agreement along the two stable branches. The main advantage of the modified conventional solution technique is its great computational efficiency which cannot be achieved by the other two strategies.

For theoretical completeness, the non-conventional solution technique is preferred as it gives the complete "story" of the power dispersion relation. However, the CPU time required by the new algorithm is much more than that required by the conventional one due to the extreme singularity of the "mass" matrix resulting from the new algorithm. As the unstable intermediate portion of the power dispersion curve[3] does not have much physical importance, the modified conventional one may be preferred due to its greater efficiency. To be both theoretically complete and computationally efficient, a combination of the two is highly recommended.

---

[3]Refer to § 7.4.

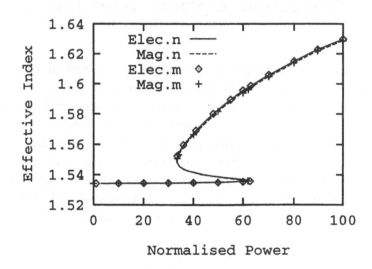

Figure 6.37: Comparison of power dispersion curves (β/k_0 *vs* $\widetilde{P}$) resulting from the modified conventional and the non-conventional solution algorithms for both the electric- and the magnetic-field formulations, where Elec and Mag denote the electric- and magnetic-field formulations and the extensions m and n denote the modified conventional and the non-conventional solution algorithm, respectively.

To conclude this section, a new non-conventional solution algorithm, valid for both the E and the H formulations and a wide class of nonlinear mechanisms, has been proposed for the computation of the complete power dispersion curve of the nonlinear structure exhibiting a jump on the dispersion curve if computed by the conventional solution technique. The jump has been successfully explained and a useful bistability phenomenon in nonlinear optical channel waveguides has been identified. It has been shown how to modify the conventional solution technique to simulate the bistability phenomenon and demonstrated that the results from the two different solution techniques and the two different formulations are mutually consistent. Moreover, some of the factors giving rise to bistability in nonlinear optical waveguides has been discussed. The structure analysed may find applications in all-optical switching and bistable devices.

6.5   Correct Modelling of Planar Nonlinear Optical Waveguides

For planar nonlinear optical waveguides simulated in quasi-2D (using one transverse spatial variable), saturating nonlinear permittivity models are not essential *mathematically* for the scalar nonlinear wave equation to possess solutions with strong nonlinear effects, because of the 1D (one-dimensional) confinement of the field. Therefore, whether the simple Kerr or the Kerr-like saturating nonlinear permittivity model is used is determined only by physical requirements. As a result, many nonlinear planar structures have been investigated and various nonlinear waveguiding effects have been predicted by quasi-2D simulations using nonlinear permittivity models with or without saturation. As discussed in our earlier work [189], in a nonlinear situation the physical world can never behave in such a way as nothing prevents the field from focussing in the second transverse direction with strong nonlinear effects. Thus, the quasi-2D formulation can be used at most to simulate weak nonlinear effects. When the self-focussing action is stronger than the diffraction effect, all stable modes must be of 2D (two-dimensional) confinement. Consequently, the results obtained by using quasi-2D simulation of strong nonlinear effects in planar structures, which have frequently appeared in the literature, *e.g.* [214, 215], [239]-[243], need re-examination.

Nonlinear planar structures have also been formulated as quasi-3D models (using two transverse spatial variables) with the weak-guidance approximation [191]. The results are correct for strong nonlinear effects but less accurate for weak nonlinear effects due to the limited extent along the dielectric interface that can be accommodated in a numerical model. As a consequence, difficulties arise in predicting the correct threshold power.

This section presents the vectorial quasi-3D modelling of nonlinear planar optical waveguides. The computed results are compared with those from the scalar quasi-2D simulation. Computational techniques and several precautionary factors in the numerical modelling are outlined. Contrary to one's expectation, the self-focussing mechanism associated with weak nonlinear effects in a nonlinear planar structure is much more complicated than that in a nonlinear channel structure and is very difficult to characterise. The proper modelling of these planar structures is fully discussed.

As an example [244], consider an isotropic thin-film nonlinear optical waveguide structure consisting of a linear high-refractive-index thin film sandwiched between two identical nonlinear layers as shown in Figure 6.38. The relative permittivity dyadic is taken to be

$$\hat{\epsilon}_r = \epsilon_r \hat{I} = \begin{cases} 1.57^2 \hat{I} & |\bar{x}| \le 6.1, \quad |\bar{y}| \le 20 \\ \left(1.55^2 + \dfrac{\|\widetilde{\boldsymbol{E}}\|_2^2}{1 + \alpha \|\widetilde{\boldsymbol{E}}\|_2^2}\right)\hat{I} & 6.1 < |\bar{x}| \le 30, \quad |\bar{y}| \le 20 \end{cases} \quad (6.30)$$

and the boundary conditions are assumed to represent perfect conductors. The saturation parameter α is again taken to be 1.0 and the normalisation factor $\rho = k_0$.

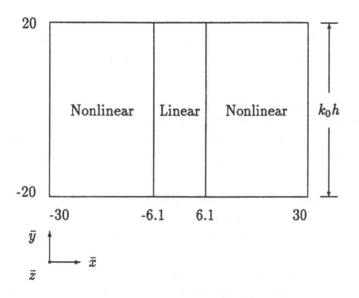

Figure 6.38: A thin-film nonlinear optical waveguide structure and coordinate system.

6.5.1   *E*-vector quasi-3D model

The whole structure is meshed into 166 second-order elements with 363 nodal points, as shown in Figure 6.39. After node renumbering, the maximum node index difference between any two related nodes is 50.

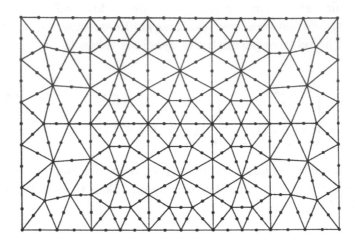

Figure 6.39: Mesh network of the thin-film nonlinear optical waveguide structure (166 second-order elements, 363 nodes).

This waveguide structure has two lines of geometrical symmetry: $\bar{x} = 0$ and $\bar{y} = 0$. In the following discussions, the terms "symmetric", "antisymmetric" and "asymmetric" are all referred to the line $\bar{x} = 0$ unless noted otherwise, as the symmetry line $\bar{y} = 0$ is trivial for a planar structure.

Quasi-3D simulation of the above nonlinear planar structure with a vectorial formulation is much more difficult than that of channel structures with only one nonlinear region. The above structure can support modes of either 1D or 2D confinement, and both types can have either a symmetric or an asymmetric form for the fundamental mode above a certain power. Also, the modal field may take the artificial boundary, $|\bar{x}| = 30$, as its symmetry plane. Another difficulty arising is that the field of 1D confinement is very unstable at high powers and may switch to a mode of 2D confinement during the iteration process. Furthermore, since the two modes with dominant electric-field component polarised in the x- and the y-direction are almost degenerate, care must be exercised in the computation. In short, special computational skills are required to facilitate the analysis.

The power dispersion curves computed by the E formulation are given in Figure 6.40. The lower portion of the dashed curve corresponds to the symmetric TE_{10} mode of 1D confinement whose dominant electric-field component is polarised in the y-direction, as shown in Figure 6.41. These solutions are identical to those obtained from a scalar quasi-2D model. At relatively low powers, the modal field corresponding to this portion is expected to be stable. At rather high powers, it is unstable, and changes to a pattern having 2D confinement. The accurate determination of the critical power at which the transition occurs strictly needs a propagation stability analysis. Also, theoretically, the lower portion of the dashed curve extends to an infinite value of $\tilde{P}$. Our computations have been performed only up to $\tilde{P} = 100$.

Modes corresponding to the upper and intermediate portions of the dashed curve are of 2D confinement. Those associated with the upper portion are expected to be stable. Figure 6.42 illustrates the modal electric field on the upper portion of the dashed curve at $\tilde{P} = 42$. Modes corresponding to the intermediate portion of the dashed curve are nonlinear surface waves and are expected to be unstable. We traced this portion only up to $\tilde{P} \approx 52.7$ (which explains the gap) as such modes are less important and also the computation of them is much more expensive than that of stable modes.

It is also possible to find the antisymmetric and the asymmetric modes of 1D confinement. We shall not go into that detail as these problems can be solved more easily by a quasi-2D model.

The solid curve in Figure 6.40 represents modes of 2D confinement whose dominant electric-field component is polarised in the x-direction. The lower portion of the solid curve corresponds to the quasi-TM_{11} mode. Similar to the strip-loaded channel structure in the previous section, it can be seen that bistability is also possible. Typical electric field patterns on the lower and the upper portions of the solid curve are shown in Figures 6.43 and 6.44.

For both power dispersion curves, the intermediate and upper portions are actually degenerate due to the geometrical symmetry of the structure as the self-focussing may occur in either nonlinear region. Obviously, the strongly self-focussed modal fields corresponding to the upper portions of the two curves can *never* be simulated by a quasi-2D model!

For the structure analysed, symmetric and antisymmetric modes (like supermodes) associated with strong nonlinear effects and confined in both transverse directions, possibly unstable, can also exist. The computation is straightforward and will be omitted for brevity.

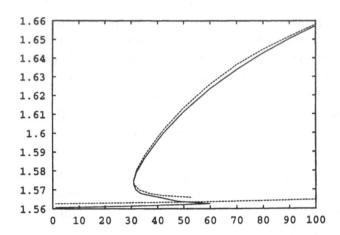

Figure 6.40: Power dispersion curves of the thin-film nonlinear optical waveguide obtained from the vector quasi-3D model: β/k_0 *vs* $\widetilde{P}$. The solid and dashed curves represent the modes whose dominant electric-field component is polarised in the x- and the y-direction, respectively. The normalised waveguide height $k_0 h = 40$.

6.5.2   Characterisation of nonlinear planar optical waveguides

In order to study the characterisation of nonlinear planar waveguides, we now present simulated results obtained from the scalar quasi-2D formulation. Here, along the $\bar{x}$-axis of the structure in Figure 6.38, 34 second-order line elements with 69 nodes are used in the finite element computation.

Considering TE waves only, when the height of the waveguide (the geometrical extent along the dielectric interface) approaches infinity it may be assumed that $\widetilde{E}_y$ depends only on $\bar{x}$. The nonlinear wave equation is given by

$$\frac{d^2 \widetilde{E}_y}{d\bar{x}^2} + \left(\epsilon_r - (\frac{\beta}{k_0})^2 \right) \widetilde{E}_y = 0 \tag{6.31}$$

and $\|\widetilde{E}\|_2^2$ in Equation 6.30 for this formulation is $|\widetilde{E}_y|^2$.

The transmitted power per unit normalised length along the $\bar{y}$-axis

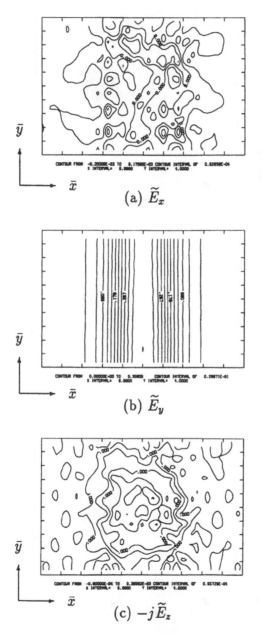

(a) $\widetilde{E}_x$

(b) $\widetilde{E}_y$

(c) $-j\widetilde{E}_z$

Figure 6.41: The normalised electric field distributions of the symmetric TE_{10} mode of 1D confinement corresponding to the lower portion of the dashed power dispersion curve in Figure 6.40 for $\widetilde{P} = 100$.

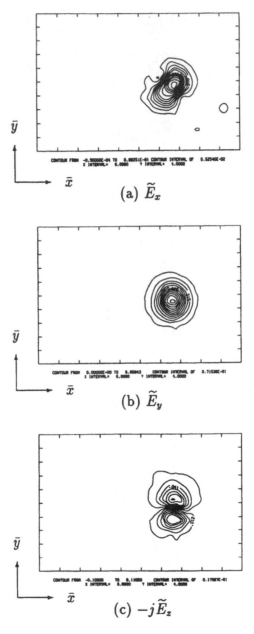

Figure 6.42: The normalised electric field distributions of the asymmetric mode of 2D confinement corresponding to the upper portion of the dashed power dispersion curve in Figure 6.40 for $\widetilde{P} = 42$.

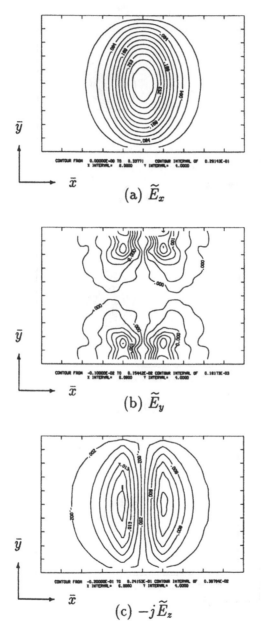

Figure 6.43: The normalised electric field distributions of the symmetric quasi-TM$_{11}$ mode of 2D confinement corresponding to the lower portion of the solid power dispersion curve in Figure 6.40 for $\widetilde{P} = 42$.

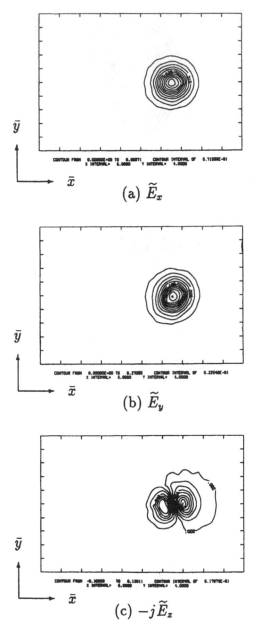

Figure 6.44: The normalised electric field distributions of the asymmetric mode of 2D confinement corresponding to the upper portion of the solid power dispersion curve in Figure 6.40 for $\widetilde{P} = 42$.

is evaluated as

$$p = \frac{1}{2Z_0 a k_0^2} \left(\frac{\beta}{k_0} \int_{-\infty}^{\infty} |\widetilde{E}_y|^2 \, d\bar{x} \right) \tag{6.32}$$

The dimensionless *normalised power per unit normalised length* defined by

$$\widetilde{p} \triangleq \begin{cases} 2Z_0 a k_0^2 p & \text{scalar quasi-2D simulation} \\ \frac{1}{k_0 h} \widetilde{P} & \textbf{\textit{E}}\text{-vector quasi-3D simulation} \end{cases} \tag{6.33}$$

will be employed to facilitate the analysis. Here $k_0 h$ is the normalised height of the waveguide. Since our coordinates are always normalised, in the following discussion it will be understood that the phrase "per unit length" means "per unit normalised length."

The nonlinear iteration procedure adopted is a scalar quasi-2D version derived from those in §6.4.

Figure 6.45 shows the computed power dispersion curve for the TE$_0$ mode. The observed all-optical switching and bistability phenomenon are already well understood. Nevertheless, we believe that this complete power dispersion relation using the scalar quasi-2D formulation has been obtained for the first time by finite element methods.

Compared to the vector quasi-3D model, the scalar quasi-2D model yields similar results for weak nonlinear effects but unphysical results for strong nonlinear effects owing to the modal field being confined in one transverse direction only. However, the vectorial quasi-3D model is much more expensive in computation. Thus it is important to compare these two models since scalar quasi-2D models have been employed extensively in simulating nonlinear waves in planar structures, including strong nonlinear effects. More importantly, we wish to determine whether nonlinear planar waveguides are better characterised by the total guided power or by the power per unit length.

The power dispersion relation resulting from a quasi-2D model is described by the effective index versus the power per unit length, whereas the one from a quasi-3D model is described by the effective index versus the total guided power. To compare these two models, it is essential to describe the latter as the effective index versus the power per unit length, namely, the normalised power divided by the normalised height of the waveguide, $k_0 h$. This converted power dispersion curve for the fundamental mode is shown in Figure 6.46 together with the one obtained from the quasi-2D model for comparison. The lower portions of the

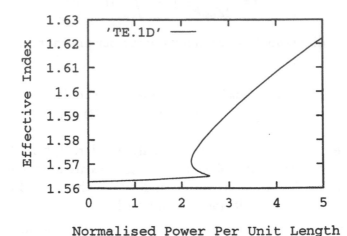

Figure 6.45: Power dispersion curve of the thin-film nonlinear optical waveguide obtained from the scalar quasi-2D model for the TE_0 mode: β/k_0 *vs* $\tilde{p}$.

two power dispersion curves are indistinguishable. However, significant discrepancy is observed in the upper portions of the two power dispersion curves solely because the modal field resulting from the quasi-2D model can have only 1D confinement.

Now it is clear that the modal fields associated with the upper portion of the power dispersion curve must be formulated in quasi-3D. Consequently, nonlinear planar waveguides operating in their strongly nonlinear regime must be characterised by the effective index versus the total guided power rather than the power per unit length. However, nothing yet can be concluded about weakly nonlinear planar waveguides since the height of the waveguide employed in the present quasi-3D model has been rather restricted due to the extensive computations involved. When the height of the waveguide is increased, it is expected that a modal field of 2D confinement associated with weak nonlinear effects will be observed in analogy with weakly-guiding linear channel waveguides. More importantly, the threshold power associated with the present quasi-3D model remains uncertain. In the following we consider solely weak nonlinear effects in order to avoid extensive computations.

When the height of the waveguide is doubled to $k_0 h = 80$, the power

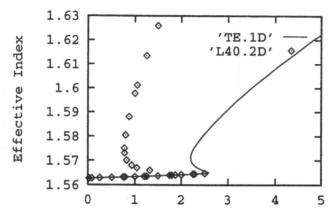

Figure 6.46: Comparison of the power dispersion curves of the thin-film nonlinear optical waveguide obtained from the vector quasi-3D ($k_0 h = 40$) and scalar quasi-2D models for the fundamental mode: β/k_0 *vs* $\tilde{p}$.

dispersion curves associated with weak nonlinear effects resulting from the **E**-vector formulation become those shown in Figure 6.47, where the solid line corresponds to the TE_{10} mode having 1D confinement while the dashed line corresponds to a new quasi-TE_{11} mode having 2D confinement. The **E** field of the TE_{10} mode is polarised in the y-direction as before and that of the quasi-TE_{11} mode has a dominant E_y component as shown in Figure 6.48 for $\tilde{P} = 80$. We believe that, in practice, the latter is more realistic than the former. This quasi-TE_{11} mode results from self-induced guidance in the nonlinear cladding. The mesh used in the computation has 184 second-order triangular elements with 401 nodes.

Note that we attempted to find this quasi-TE_{11} mode when $k_0 h = 40$ without success, indicating that the height of the structure originally chosen is insufficient for the waveguide to support this quasi-TE_{11} mode. Thus care must be exercised in choosing the height of a planar waveguide in a quasi-3D model.

Next, we try to compare the threshold power resulting from the two simulations, namely, the scalar quasi-2D simulation and the **E**-vector quasi-3D simulation when $k_0 h = 80$. Figure 6.49 shows the power dis-

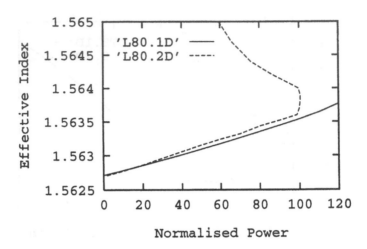

Figure 6.47: Power dispersion curves of the thin-film nonlinear optical waveguide obtained from the vector quasi-3D model when $k_0 h = 80$: β/k_0 *vs* $\widetilde{P}$. The solid and dashed curves represent modes whose electric fields have 1D and 2D confinement, respectively.

persion curves of the scalar TE_0 mode and the vector quasi-TE_{11} mode in terms of the normalised power per unit length. Apparently the lower portions of the two power dispersion curves are very close, indicating that weak nonlinear effects in a planar waveguide are characterised by the power per unit length rather than the total guided power. However, the quasi-3D model has a much lower threshold power per unit length than the quasi-2D model solely because the former allows the modal field to confine in both transverse directions, which is realistic practically.

Now it remains to answer how the threshold power is affected by the height of the waveguide used in a quasi-3D model. Specifically, we ask whether the threshold power and the threshold power per unit length increase or decrease with increasing height of the waveguide. In order to avoid further increasing the height of the waveguide so as to minimise the computational cost, we consider the quasi-TM_{11} mode, whose dominant E-field component is polarised in the x-direction. The threshold power of this mode can be obtained numerically for a normalised waveguide height of either 40 or 80. This mode corresponds to the TM_0 mode from a quasi-2D model. Figure 6.50 illustrates the relevant portions of the

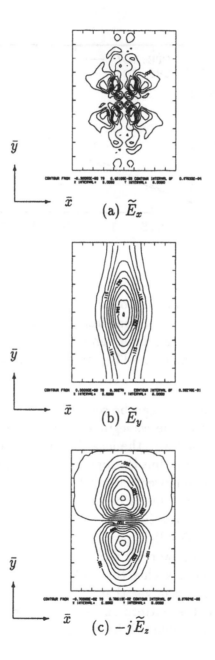

Figure 6.48: The normalised electric field distributions of the quasi-TE$_{11}$ mode with 2D confinement corresponding to the lower portion of the dashed power dispersion curve in Figure 6.47 for $\widetilde{P} = 80$ when $k_0 h = 80$.

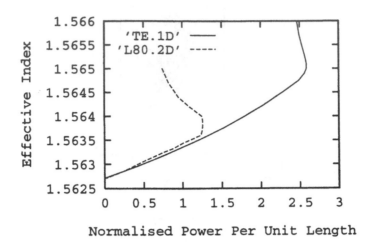

Figure 6.49: Comparison of the power dispersion curves of the thin-film nonlinear optical waveguide resulting from the scalar quasi-2D simulation (solid curve showing the TE_0 mode) and the vector quasi-3D simulation (dashed curve showing the quasi-TE_{11} mode) when $k_0 h = 80$: β / k_0 vs $\tilde{p}$.

power dispersion curves. As can be seen, the threshold power increases with increasing height of the waveguide. This is understandable as the taller the waveguide, the lower the local intensity of the electric field for the same guided power and thus the higher the threshold power. It can be shown that the threshold power per unit length decreases with increasing height of the waveguide. This is because, for the same power per unit length, the taller the waveguide, the more the total guided power and thus the stronger the 2D self-focussing action. These results show that neither the total guided power nor the power per unit length is a suitable parameter to characterise such a threshold. In practice, the height of a planar waveguide is much larger than the diameter of a launched laser beam. Consequently, the pattern size of the laser beam end-fired into a nonlinear planar waveguide will have a great bearing on the threshold power.

One might have noticed that there exists some discrepancy between the effective indices of the two power dispersion curves in Figure 6.50 at very low powers. This discrepancy indicates that the heights of the waveguide used in the numerical example are inadequate to compute

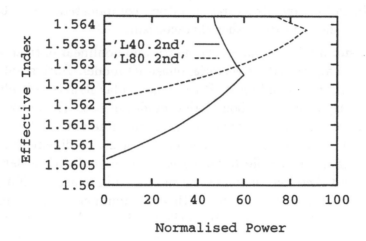

Figure 6.50: Comparison of the threshold powers of the thin-film nonlinear optical waveguide obtained from the vector quasi-3D simulation for the quasi-TM_{11} mode with a normalised waveguide height of 40 (solid curve) and 80 (dashed curve): β/k_0 vs $\tilde{P}$.

the quasi-TM_{11} mode in a nearly linear situation. Being the lowest TM mode, it corresponds to the TM_0 mode of the infinitely-extending planar structures, and the TM_0 mode can be easily found by means of a scalar quasi-2D model which is less expensive and more accurate than a vector quasi-3D model with limited extent in the direction along the dielectric interface. Thus it is advantageous to use a quasi-2D model to simulate very weak nonlinear effects in a planar structure.

It should be mentioned that with models in quasi-3D, all modes which can be found by seeking the effective index for a given guided power are expected to be stable[4]. However, this is not true for models in quasi-2D as a mode which is stable with a quasi-2D simulation may not be stable with a quasi-3D simulation. This will be a disadvantage of quasi-2D models since propagation stability analyses would additionally need to be performed.

---

[4]Symmetric and antisymmetric modes, particularly those associated with strong nonlinear effects, could be exceptions if they are computed by making use of the geometrical symmetry of the structure.

In short, the simulation of the nonlinear quasi-TE and quasi-TM modes corresponding to the lower portions of the power dispersion curves of nonlinear planar structures is rather complicated. The bistability power threshold needs further characterising.

To summarise, a planar, symmetric and nonlinear optical waveguide has been analysed with a vectorial quasi-3D model. Both stable and unstable solutions are obtained. For the particular structure studied which has perfectly conducting boundaries, modes of both 1D and 2D confinement can exist. The former are identical to those obtained with a scalar quasi-2D model, as expected. The modes of 2D confinement associated with strong nonlinear effects can never be simulated with scalar quasi-2D models, and therefore they must be simulated in quasi-3D. Furthermore, the lowest TE mode associated with weak nonlinear effects can have either a 1D or 2D confinement. It is believed that the latter (the quasi-TE_{11} mode) is more realistic than the former. However, the latter cannot be found unless the geometrical extent along the dielectric interface of the planar structure is sufficiently large in the vector quasi-3D model.

It has been shown that strong nonlinear effects in a planar structure are characterised by the total guided power, whereas the weak nonlinear effects are approximately characterised by the power per unit length. The scalar quasi-2D simulation, which is employed extensively by researchers, is accurate only for very weak nonlinear effects. The rather limited results presented here indicate that the associated bistability threshold is characterised by neither the total guided power nor the power per unit length. In practice, the size of the laser beam launched into a planar nonlinear structure may contribute a great deal to its threshold.

Chapter 7

Weak-Guidance Approximation and Propagation Stability Analysis

7.1   Introduction

As is known, the computation of nonlinear waveguide problems is much more expensive than that of their linear counterparts. Therefore, it is of practical importance to simplify the computation by making reasonable approximations where appropriate. In analysing linear optical waveguides supporting only one or two modes, the scalar or weak-guidance approximation may be applied by solving the *scalar Helmholtz* wave equation instead of the *curl-curl* equation. As nonlinear optical waveguides behave so differently from linear ones, the feasibility of applying the scalar approximation method to nonlinear structures should be carefully examined. Though the scalar approximation method has been applied to quasi-3D nonlinear planar structures by Akhmediev *at al.* [190]-[192], the applicability of such a method to other quasi-3D structures, such as nonlinear channel waveguides, has remained unknown. Most importantly, such an approximation method should be justified by comparing its results with those from the rigorous vectorial formulation. We have already reported some preliminary results of a comparison of the scalar and vector FEMs applied to a nonlinear planar structure [189]. Here we will extend the comparison to other structures, including a nonlinear strip-loaded channel waveguide.

Asymmetric planar nonlinear optical waveguides have been analysed in quasi-2D and some interesting features have been predicted [199,
200]. For strong nonlinear effects, however, planar nonlinear structures
must be formulated in quasi-3D to simulate localised self-focussing, as
discussed in § 6.5. Here an asymmetric planar nonlinear optical waveguide is analysed in quasi-3D with an inexpensive procedure: the scalar
approximation.

Up to this stage we have developed a systematic procedure for the
modal analysis of nonlinear optical waveguides. For such a procedure
to make sense, however, at least some of the predicted nonlinear modes
must be stable. In other words, nonlinear guided waves could only be
investigated by propagation methods if all nonlinear modes happened
to be unstable. Therefore, the stability analysis of nonlinear modes is
crucial to the application of nonlinear modal methods.

From the author's experience, as discussed previously, all modes
which can be found by seeking the effective index for a given guided
power are expected to be stable provided that the whole structure is
modelled. However, this should be confirmed by a propagation stability
analysis.

If stable nonlinear modes exist, the propagation of a stable nonlinear
mode will be similar to that of a linear mode in the sense that both can
keep their profiles during propagation without dispersing or focussing,
which is a desirable feature in applications. The difference between linear
and stable nonlinear modes is that nonlinear modes are not subject to
linear superposition; therefore the coupling of nonlinear modes should
be investigated by nonlinear coupled mode theories [245]-[258] and/or
nonlinear wave/beam propagation techniques [259]-[266].

The accurate analytical approach to the stability analysis of nonlinear modes is almost intractable in the case of quasi-3D structures. We
shall adopt a numerical approach to perform such an analysis, as was
done in [243] and [267]-[270] for quasi-2D structures.

The numerical computation of the propagation of vectorial waves in
a 3D nonlinear waveguide structure for several hundred steps is daunting
and perhaps a supercomputer is needed to perform such an analysis.
Here, we confine ourselves to scalar wave propagation, which is believed
to be sufficient for the study of the propagation stability of modal fields
in those structures to which scalar modal analysis is suited.

The scalar nonlinear wave propagation in the present work is achieved by the *finite-element plus finite-difference method* (FEM-FDM), which is a generalisation of the work in [271] from *2D linear* structures to *3D nonlinear* structures and is similar in some way to the work in [266]. The FEM-FDM technique has several advantages over both the conventional *beam propagation method* (BPM) [272]-[278] based on the *fast Fourier transform* (FFT) and the *finite-difference plus beam propagation method* (FD-BPM) [279], and we refer to [266, 271] for details.

Beam, wave and pulse propagation in nonlinear waveguide structures has been investigated by many workers, *e.g.* [241, 242], [259]-[266], [280]-[290]. Also, the stability of nonlinear modes/waves/beams in quasi-2D structures as well as in quasi-3D structures of cylindrical symmetry has been reported [243], [266]-[270], [291]-[296]. The propagation stability analysis of nonlinear modes in general quasi-3D structures was reported in [297]. The work reported in §7.4 is devoted to justifying the method of modal analysis for nonlinear optical waveguides.

Here we are solely concerned about the propagation stability of nonlinear modes. However, the methodology and software developed can be easily generalised to cover linear/nonlinear coupled waveguides, which are investigated in Chapter 8, as well as tapered waveguides by imposing an absorbing or an infinite element boundary.

7.2 Weak-Guidance Approximation

For simplicity, we consider only non-dissipative isotropic waveguides and solve for the electric field.

Under the weak-guidance approximation for modal analysis, with the phase factor $\exp\{j(\omega t - \bar{\beta}\bar{z})\}$ being implied, each of the two transverse components of the electric field in the Cartesian coordinate system satisfy the same *scalar Helmholtz* wave equation in quasi-homogeneous isotropic media:

$$-\frac{\partial^2 E}{\partial \bar{x}^2} - \frac{\partial^2 E}{\partial \bar{y}^2} + (\bar{\beta}^2 - \epsilon_r)E = 0, \qquad (7.1)$$

where $\bar{x} = k_0 x$, $\bar{y} = k_0 y$ and $\bar{\beta} = \beta/k_0$ (that is, the normalisation used throughout this chapter corresponds to $\rho = k_0$). The discontinuity and boundary conditions for the two components may be different. The real scalar field variable E of Equation 7.1 in its strong sense must be assumed to be continuous everywhere despite the fact that there may be a jump in the permittivity. Otherwise, Equation 7.1 has no solution [8, pp. 641]. With the intermediate form of the weak formulation of Equation 7.1, a discontinuous variable E may be introduced to account for any jump in the permittivity. However, this makes little practical difference as the jump in permittivity, if any, is usually relatively small under the weak-guidance approximation. Therefore, the discontinuity of E is generally ignored.

The modes of weakly-guiding optical waveguides are approximately TEM (transverse electromagnetic) waves [8, pp. 283]. Therefore, the longitudinal component of the field is negligible. If the transverse component sought with the scalar approximation is the same as the dominant (transverse) component of the vectorial modal field, it is expected that the scalar solution will be very close to the exact one, even in nonlinear cases. The question is whether the weak-guidance approximation is also "good enough" for the computation of a modal field whose two transverse components are comparable, especially in a nonlinear situation where each transverse component represents a substantial contribution to the guided power!

Under the weak-guidance approximation, the discontinuity of the field can be reasonably ignored. And if the modal field is far from the boundary or the boundary condition is the same for the two transverse components, such as a vanishing boundary condition, the results obtained by solving the two transverse components individually are indistinguishable. Consequently, the weak-guidance approximation will yield very accurate results in those situations, at least as far as linear problems are concerned, except that the polarisation orientation of the field is by no means known. Turning to nonlinear problems, the scalar approximation will also yield very good results provided that

- the nonlinear permittivity depends on the magnitude of the electric field rather than on each component differently;

- the scalar field obtained is interpreted as the one and only component of a one-component vector field, linearly-polarised in a certain transverse direction, rather than as just one component of a multi-component vector field, for the purpose of determining the guided power.

In situations beyond those outlined above, the scalar approximation should be applied to nonlinear structures only with care.

For many useful optical waveguide devices, the weakly-guiding condition is normally satisfied. The problem is that usually it is not known *a priori* whether the modal field has a dominant (transverse) component and if so what the direction of its linear polarisation is. To overcome such difficulties, the following restrictions are imposed:

- The jump in permittivity, if any, is relatively small so that the discontinuity of the modal electric field can be ignored;

- The nonlinearity depends only on the magnitude of the electric field;

- The media involved are isotropic;

- The modal field pattern is far from the boundary or else the two transverse Cartesian components are required to satisfy the same boundary condition.

Under the above conditions, it has been the author's experience that the weak-guidance approximation can be applied to nonlinear optical waveguides with confidence.

With the scalar field being interpreted as a vector field linearly-polarised in a certain transverse direction, the normalised guided power is given by

$$\widetilde{P} \triangleq 2Z_0 a k_0^2 P = \frac{\beta}{k_0} \cdot \int_\Omega |\widetilde{E}|^2 \, d\bar{x} d\bar{y}, \tag{7.2}$$

where

$$\widetilde{E} \triangleq \sqrt{a} E \tag{7.3}$$

and the definitions of the other symbols are the same as or equivalent to those for the vectorial formulation.

Equation 7.1 is then formulated in a weak form, and application of the Galerkin FEM to the intermediate form of the weak formulation leads to a generalised matrix eigenvalue problem, the derivation being similar to but much simpler than the vectorial problem and therefore omitted for brevity. Then the eigenpair $(\bar{\beta}^2, \widetilde{E})$ can be solved by techniques similar to those used for vectorial problems for a given guided power $\widetilde{P}$.

Though all stable solutions can be found by the procedure described above, one may be interested in unstable solutions as well. To obtain complete power dispersion relations, the above procedure should be changed so that the computation can be performed by seeking the guided power for a given effective index $\bar{\beta}$. The non-conventional nonlinear iteration procedure is given by

$$-\frac{\partial^2 E'_{(i+1)}}{\partial \bar{x}^2} - \frac{\partial^2 E'_{(i+1)}}{\partial \bar{y}^2} + (\bar{\beta}^2 - \epsilon_r^l)E'_{(i+1)} = A^2_{(i+1)}\{\frac{\epsilon_{r(i)}^n}{A^2_{(i)}}\}E'_{(i+1)}, \quad (7.4)$$

where a new scalar field variable E' and a real scalar coefficient A are defined by

$$\widetilde{E} \triangleq AE', \quad (7.5)$$

subject to

$$\frac{\beta}{k_0} \cdot \int_\Omega |E'|^2 \, d\bar{x}d\bar{y} \equiv 1 \quad (7.6)$$

so that

$$\widetilde{P} \equiv A^2. \quad (7.7)$$

Hence the eigenpair sought is (A^2, E') for a given $\bar{\beta}$.

It should be mentioned that, after this work had been performed, we found that the above non-conventional nonlinear iteration procedure for the scalar Helmholtz nonlinear wave equation is, to some extent, similar to the one used in [190] for nonlinear surface waves.

To get a quantitative impression of the scalar approximation, two examples are given in the following.

7.2.1   An example of a nonlinear strip-loaded channel structure

The example chosen is the nonlinear strip-loaded channel structure given in § 6.4 (the dyadic permittivity is actually isotropic). The mesh used here is obtained by dividing each second-order element of the mesh in § 6.4 into four first-order elements, as is done in Chapter 4; and the maximum node index difference within any first-order element is now 25. Consequently, the number of unknowns will be exactly one-third of the number used in § 6.4 and the bandwidth of the global matrices will be only about one-sixth. Thus, the computation time will be greatly reduced.

For the present example, the fundamental modal electric field has a dominant component, which is E_x, as shown in § 6.4. If we solve for E_x by the scalar approximation with the associated perfectly conducting boundary condition, very accurate results are expected. Now suppose we do not know this information. We only make use of the fact that the guided region is far from the boundary, which enables us to use the vanishing boundary condition for a scalar solution. Furthermore the nonlinear permittivity is isotropic. Of course, the discontinuity of the field has to be, whether reasonably or not, ignored as the polarisation direction of the field is unknown. Under these conditions we wish to examine the accuracy of the scalar approximation.

A quantitative comparison of the power dispersion curves resulting from the scalar and the vector solution is given in Figure 7.1, showing good agreement between the two formulations. On the upper branch of the curve, the discrepancy does increase with increasing power. This is understandable as the mesh used is rather coarse for such a strongly self-focussed field and different orders of elements were employed albeit with the same total number of nodes.

As the power dispersion relation resulting from the scalar approximation is so close to that resulting from the rigorous vector formulation, the magnitude of the normalised scalar electric field, $|\widetilde{E}|$, is expected to be very similar to the magnitude of the normalised electric field from the corresponding vector formulation, $\|\widetilde{\boldsymbol{E}}\|_2$. These are plotted in Figures 7.2 to 7.4 for points on the three different branches of the power dispersion curve at almost the same guided power for comparison.

7.2.2 An example of a symmetric planar nonlinear structure

The example chosen is the symmetric planar nonlinear structure given in § 6.5 (the dyadic permittivity is actually isotropic) with $k_0 h = 40$. The mesh employed is obtained in the same way as in the previous subsection. Here the maximum node index difference within any first-order element is 25.

For this example, modal fields of 1D confinement can exist, as shown in § 6.5; and those modal fields can be approximated by the single component E_y with perfectly conducting boundary conditions. Here we are concerned only with modes of 2D confinement. The computation is performed in the same way as in the previous subsection, *i.e.* by using a

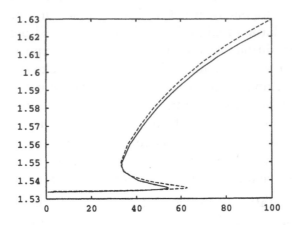

Figure 7.1: A comparison of the power dispersion curves (β/k_0 *vs* $\widetilde{P}$) resulting from the scalar (solid line) and the vector (dashed line) formulation for the nonlinear strip-loaded channel structure.

continuous real scalar field with vanishing boundary conditions.

A comparison of the power dispersion curves resulting from the scalar and the vector solution of the modal field with 2D confinement is given in Figure 7.5, where the vector solution corresponds to the mode whose dominant electric-field component is polarised in the x-direction (quasi-TM$_{11}$ mode). As can be seen, the results from the two formulations are in excellent agreement. The intersection of the two curves at $\widetilde{P} \approx 90$ is understandable. As the nonlinear region is rather large compared with the size of the field pattern at high power, the self-focussing can take place at different areas of the nonlinear region. However, different areas are associated with slightly different fineness of mesh. Consequently, there is a range of uncertainty of the results obtained due to the rather coarse mesh adopted.

Comparisons of the field distributions resulting from the vector and the scalar formulations for $\widetilde{P} = 42$ on the two stable branches are given in Figures 7.6 and 7.7. Since the structure is symmetric, the modal field can be self-focussed in either nonlinear region. The field patterns in Figure 7.7 reflect this.

Summarising this section, a comparison of the scalar and the vector formulation of nonlinear optical waveguides has been made for both pla-

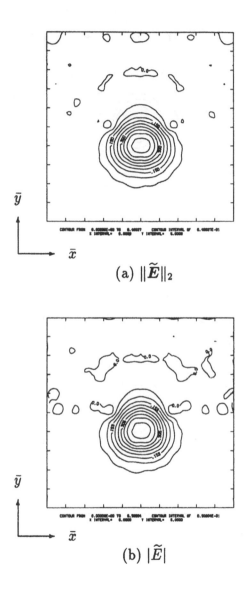

(a) $\|\widetilde{\boldsymbol{E}}\|_2$

(b) $|\widetilde{E}|$

Figure 7.2: A comparison of the magnitude distributions of the normalised electric field resulting from (a) the vector and (b) the scalar formulation on the lower branch of the power dispersion curve in Figure 7.1 for $\widetilde{P} = 44$.

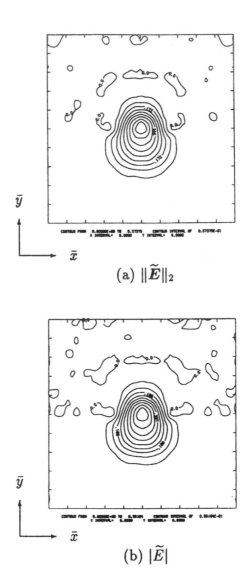

(a) $\|\widetilde{\boldsymbol{E}}\|_2$

(b) $|\widetilde{\boldsymbol{E}}|$

Figure 7.3: A comparison of the magnitude distributions of the normalised electric field resulting from (a) the vector and (b) the scalar formulation on the middle branch of the power dispersion curve in Figure 7.1 for $\widetilde{P} \approx 44$.

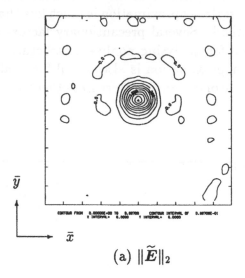

(a) $\|\widetilde{\boldsymbol{E}}\|_2$

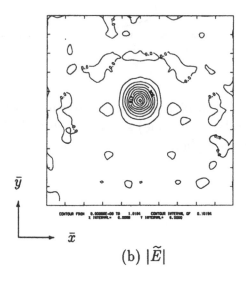

(b) $|\widetilde{E}|$

Figure 7.4: A comparison of the magnitude distributions of the normalised electric field resulting from (a) the vector and (b) the scalar formulation on the upper branch of the power dispersion curve in Figure 7.1 for $\widetilde{P} = 44$.

nar and channel structures, and excellent agreement has been achieved
due to the weakly-guiding nature of the structures. However, the CPU
time required by the scalar approximation is much less than that required
by the rigorous method. Several precautionary factors in utilising the
scalar approximation for analysing nonlinear structures have been discussed. The scalar approximation is also useful for qualitative analysis
and for the initial stage of design for structures other than weakly-guiding
structures.

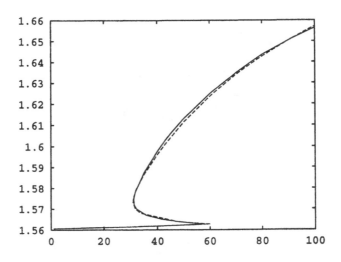

Figure 7.5: A comparison of the power dispersion curves (β/k_0 *vs* $\widetilde{P}$)
resulting from the scalar (solid line) and the vector (dashed line) formulation for the symmetric planar nonlinear structure (quasi-TM_{11} mode
at low power).

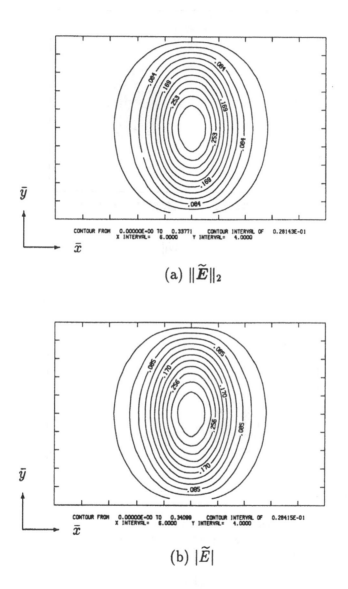

(a) $\|\widetilde{\boldsymbol{E}}\|_2$

(b) $|\widetilde{E}|$

Figure 7.6: A comparison of the magnitude distributions of the normalised electric field resulting from (a) the vector and (b) the scalar formulation on the lower branch of the power dispersion curve in Figure 7.5 for $\widetilde{P} = 42$.

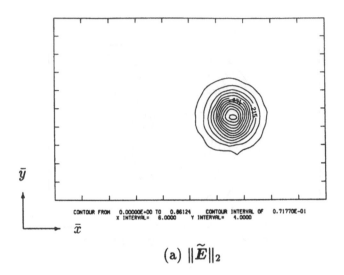

CONTOUR FROM   0.00000E+00 TO   0.86124   CONTOUR INTERVAL OF   0.71770E-01
X INTERVAL=   6.0000   Y INTERVAL=   4.0000

(a) $\|\widetilde{\boldsymbol{E}}\|_2$

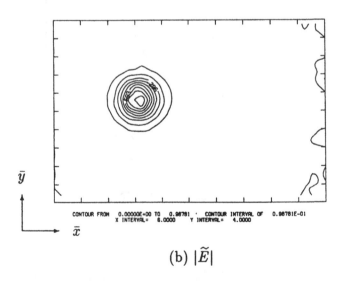

CONTOUR FROM   0.00000E+00 TO   0.98781   CONTOUR INTERVAL OF   0.98781E-01
X INTERVAL=   6.0000   Y INTERVAL=   4.0000

(b) $|\widetilde{E}|$

Figure 7.7: A comparison of the magnitude distributions of the normalised electric field resulting from (a) the vector and (b) the scalar formulation on the upper branch of the power dispersion curve in Figure 7.5 for $\widetilde{P} = 42$.

7.3 Weak-Guidance Approximation for an Asymmetric Planar Nonlinear Structure

The asymmetric planar nonlinear optical waveguide to be analysed is the one given in the previous section, except that the nonlinear coefficient for the nonlinear region on the right-hand side is halved. Now the field and the power are normalised with respect to the higher nonlinear coefficient on the left-hand side. The same mesh is used and only the modal electric field with 2D confinement is sought using the scalar approximation.

The power dispersion relation is given in Figure 7.8. There are now two unconnected curves. The left-hand curve is similar to that of the symmetric structure given in the previous section. Typical electric field distributions on this curve are shown in Figures 7.9 to 7.11, and are similar to those of the symmetric structure, exhibiting the field gradually diffusing and then sharply switching from the linear film into the bounding medium of higher nonlinear coefficient with increasing power.

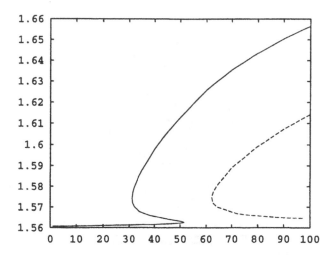

Figure 7.8: The power dispersion relation (β/k_0 *vs* $\widetilde{P}$) for the asymmetric planar nonlinear structure.

The right-hand curve exists only above a certain guided power. Its upper branch corresponds to field localisation in the bounding medium

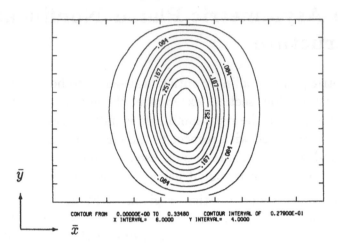

Figure 7.9: The normalised electric field distribution resulting from the scalar formulation on the lower branch of the left-hand curve of the power dispersion relation in Figure 7.8 for $\widetilde{P} = 40$.

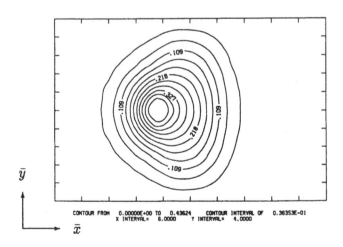

Figure 7.10: The normalised electric field distribution resulting from the scalar formulation on the middle branch of the left-hand curve of the power dispersion relation in Figure 7.8 for $\widetilde{P} \approx 40$: a nonlinear surface wave.

of smaller nonlinear coefficient; its lower branch corresponds to a single interface nonlinear surface wave. Typical field distributions for this curve are shown in Figures 7.12 and 7.13. We have performed special numerical simulations, and the results show that the nonlinear surface wave is unstable and always switches to the stable solution on the same curve having the same power. Obviously, when the guided power is lower than the threshold power of this curve, the field will switch to one corresponding to the left-hand curve. Hence this structure could possibly be used for all-optical switching and/or low-threshold devices.

It is also possible to find modes in which self-focussing occurs in both nonlinear regions simultaneously. We shall not involve ourselves in this work as the immediate application of such a phenomenon is not obvious.

To conclude this section, an asymmetric planar nonlinear structure has been simulated in quasi-3D with the scalar approximation. The power dispersion relation obtained has two unconnected curves. One is similar to those of the symmetric planar and the strip-loaded channel structures, where switching and bistability are possible. The other exists only above a certain power, showing switching and low-threshold effects. Qualitatively, these phenomena are somewhat similar to those obtained from the quasi-2D simulation without saturation reported in [200]. In practice, however, a modal field of 1D confinement is unstable for strong nonlinear effects, whereas a modal field of 2D confinement on the upper branches of the power dispersion curves is expected to be stable. The structure investigated may find applications in all-optical switching, low-threshold and bistable or even multistable devices.

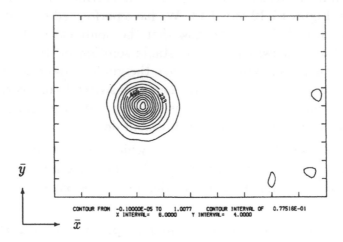

Figure 7.11: The normalised electric field distribution resulting from the scalar formulation on the upper branch of the left-hand curve of the power dispersion relation in Figure 7.8 for $\widetilde{P} = 40$.

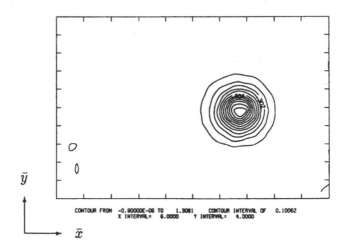

Figure 7.12: The normalised electric field distribution resulting from the scalar formulation on the upper branch of the right-hand curve of the power dispersion relation in Figure 7.8 for $\widetilde{P} = 80$.

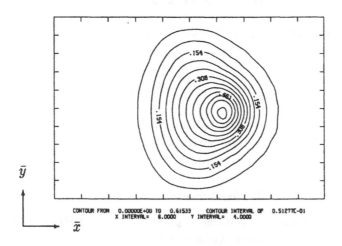

Figure 7.13: The normalised electric field distribution resulting from the scalar formulation on the lower branch of the right-hand curve of the power dispersion relation in Figure 7.8 for $\widetilde{P} \approx 100$: a nonlinear surface wave.

7.4   Nonlinear Wave Propagation in 3D based on FEM-FDM and the Stability Analysis of Nonlinear Modes

The propagation of scalar waves is performed by the finite-element plus finite-difference method (FEM-FDM) [266, 271], in which the transverse dependence is approximated by the finite element method, and the longitudinal dependence is approximated by the finite difference method.

7.4.1   Propagation algorithm

The scalar wave equation for 3D weakly-guiding optical structures with an assumed time-dependence of $\exp\{j\omega t\}$ is given by

$$(\frac{\partial^2}{\partial \bar{x}^2} + \frac{\partial^2}{\partial \bar{y}^2} + \frac{\partial^2}{\partial \bar{z}^2})E + \epsilon_r E = 0, \qquad (7.8)$$

where the coordinates are normalised with respect to k_0. The complex scalar electric field E can be written in the form of a slowly-varying amplitude multiplied by a phase factor:

$$E(\bar{x}, \bar{y}, \bar{z}) \triangleq \psi(\bar{x}, \bar{y}, \bar{z})e^{-j\tilde{\beta}\bar{z}}, \qquad (7.9)$$

where $\tilde{\beta}$ is a normalised reference propagation constant and ψ is the slowly-varying complex amplitude function.

Substituting Equation 7.9 into Equation 7.8 and assuming that

$$|\frac{\partial^2 \psi}{\partial \bar{z}^2}| \ll \tilde{\beta}|\frac{\partial \psi}{\partial \bar{z}}| \qquad (7.10)$$

so that the second-order derivative with respect to $\bar{z}$ can be neglected, the following parabolic wave equation is obtained:

$$j2\tilde{\beta}\frac{\partial \psi}{\partial \bar{z}} = \frac{\partial^2 \psi}{\partial \bar{x}^2} + \frac{\partial^2 \psi}{\partial \bar{y}^2} + (\epsilon_r - \tilde{\beta}^2)\psi. \qquad (7.11)$$

Equation 7.11 is then formulated in a weak form with respect to the transverse coordinates, as is done in Chapter 2. Application of the Galerkin FEM to the intermediate form of the weak formulation of Equation 7.11 in the transverse cross-section yields the following matrix equation:

$$j2\tilde{\beta}A\frac{\partial \psi}{\partial \bar{z}} = (B + C)\psi, \qquad (7.12)$$

where A and B are real constant matrices in general, but matrix C and the column vector of the nodal unknowns ψ may be complex functions of $\bar{z}$. C is also a function of ψ in the nonlinear case. If $\{w_k\}$ and $\{v_l\}$ are sets of testing and expansion functions, respectively, the elements of matrices A, B and C are given by

$$A_{kl} = \int_\Omega w_k v_l \, d\bar{x}d\bar{y},$$

$$B_{kl} = -\int_\Omega \left(\frac{\partial w_k}{\partial \bar{x}}\frac{\partial v_l}{\partial \bar{x}} + \frac{\partial w_k}{\partial \bar{y}}\frac{\partial v_l}{\partial \bar{y}}\right) d\bar{x}d\bar{y},$$

$$C_{kl} = \int_\Omega (\epsilon_r - \tilde{\beta}^2)w_k v_l \, d\bar{x}d\bar{y}. \qquad (7.13)$$

Here Ω is the cross-section of the waveguide, and w_k and v_l are assumed to be real.

Numerically integrating Equation 7.12 from point $\bar{z}_n$ to point $\bar{z}_{n+1} \overset{\triangle}{=} \bar{z}_n + \Delta\bar{z}$ by means of the trapezoidal rule and rearranging the order, the following matrix equation is obtained:

$$S_{n+1}\psi_{n+1} = T_n\psi_n; \qquad n = 0, 1, 2, \cdots \qquad (7.14)$$

where

$$\psi_n \overset{\triangle}{=} \psi(\bar{z}_n), \qquad (7.15)$$

and

$$S_n = j2\tilde{\beta}A - 0.5\Delta\bar{z}(B + C(\bar{z}_n)), \qquad (7.16)$$

and

$$T_n = j2\tilde{\beta}A + 0.5\Delta\bar{z}(B + C(\bar{z}_n)). \qquad (7.17)$$

The given initial condition ψ_0 consists of the nodal values of the complex scalar electric field distribution to be propagated.

Given the fact that

$$T_n = -S_n^*, \qquad (7.18)$$

only matrix C needs to be calculated at each step of the propagation. In the nonlinear case, however, iterations are required to evaluate the nonlinear term in C.

Equation 7.14 is a deterministic equation with complex and banded matrices. It can be solved by a standard subroutine in the *International Mathematical and Statistical Libraries* (IMSL).

Based on the above algorithm, a linear/nonlinear scalar wave propagation program has been developed, which can perform the propagation of an initial field distribution or the continuation of previous propagation. For the purpose of plotting, the program offers the user the flexibility of choosing the plotting in either the $\bar{x}\bar{z}$-plane or the $\bar{y}\bar{z}$-plane by means of integrating the wave along the $\bar{y}$-axis or the $\bar{x}$-axis or by simply using a fixed $\bar{y}$ or $\bar{x}$ value.

It should be mentioned that the vanishing Dirichlet boundary condition has been assumed in developing the program, which is sufficient for the stability analysis of nonlinear modes. As for other applications in guided optics, infinite elements or absorbing boundaries may be required if the wave is reflected at this artificial boundary.

7.4.2 Stability analysis of nonlinear modes

The structure chosen for illustration is the nonlinear strip-loaded channel optical waveguide given in § 6.4 and § 7.2.1. The modal electric field is obtained by the scalar approximation as is done in the previous section. Again, the mesh used here is obtained by dividing each second-order element of the mesh in § 6.4 into four first-order elements. In the following, the plotting of the wave propagation is given in the $\bar{y}\bar{z}$-plane for a fixed value of $\bar{x}$, namely $\bar{x} = 0$, referring to the coordinate system given in Figure 6.27.

Figures 7.14-7.16 show the propagation properties of the nonlinear modes corresponding to the lower, the middle and the upper branches of the solid power dispersion curve in Figure 7.1, resulting from the scalar formulation, at $\widetilde{P} = 44$. As expected, the modes associated with the lower and the upper branches of the power dispersion curve are stable, whereas the mode associated with the middle branch is unstable. Thus, the bistabilty phenomenon predicted in § 6.4 is confirmed. Here, the reference propagation constant $\tilde{\beta}$ is chosen to be very close to the one calculated by the modal method, which allows us to use a rather large step size $\Delta\bar{z}$ for the propagation. The electric field distributions before and after the propagation are almost indistinguishable for the stable modes, and their plots are omitted for brevity. The unstable mode is plotted in Figure 7.17.

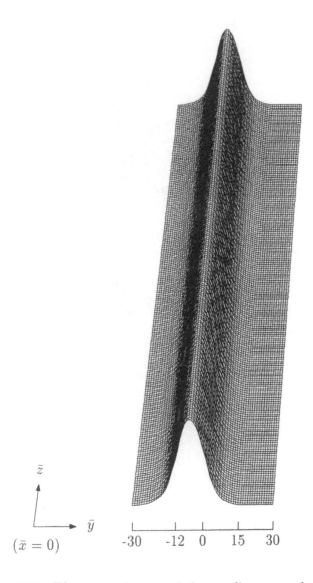

Figure 7.14: The propagation of the nonlinear mode corresponding to the lower branch of the solid dispersion curve in Figure 7.1 at $\widetilde{P} = 44$, for 200 steps with $\Delta \bar{z} = 10$ and $\widetilde{\bar{\beta}} = 1.534$.

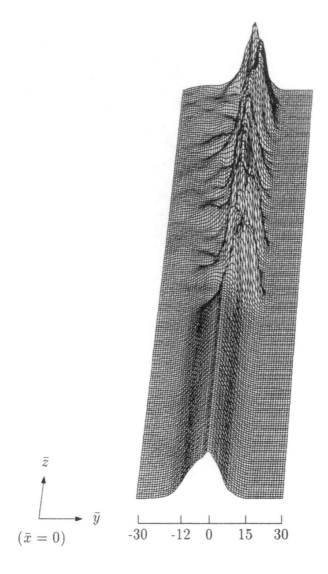

$\bar{z}$

$\bar{y}$

$(\bar{x} = 0)$

-30   -12   0   15   30

Figure 7.15: The propagation of the nonlinear mode corresponding to the middle branch of the solid dispersion curve in Figure 7.1 at $\widetilde{P} \approx 44$, for 200 steps with $\Delta\bar{z} = 10$ and $\tilde{\beta} = 1.539$.

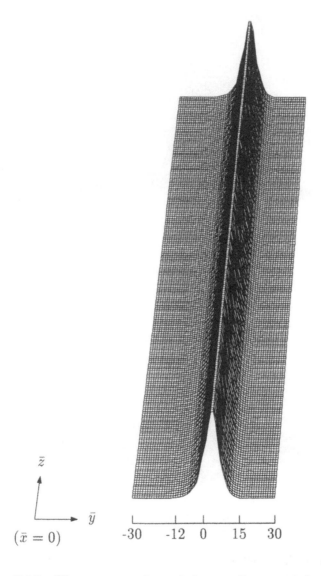

$\bar{z}$

$\bar{y}$

$(\bar{x} = 0)$

-30   -12   0   15   30

Figure 7.16: The propagation of the nonlinear mode corresponding to the upper branch of the solid dispersion curve in Figure 7.1 at $\widetilde{P} = 44$, for 200 steps with $\Delta\bar{z} = 10$ and $\tilde{\beta} = 1.58$.

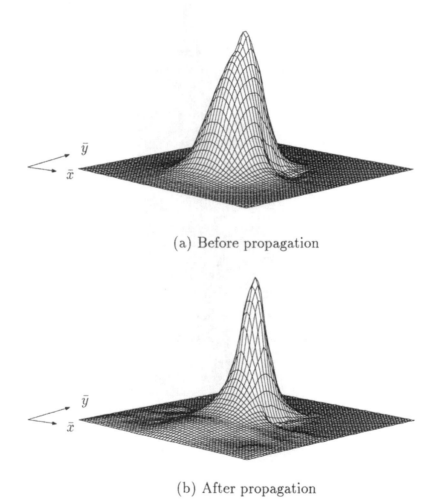

(a) Before propagation

(b) After propagation

Figure 7.17: The scalar electric field distributions before and after propagation for the unstable mode whose propagation is shown in Figure 7.15.

7.4.3 Propagation behaviour of nonlinear quasi-modes

Nonlinear modes are spatial solitons [298]-[305]. Fundamental nonlinear modes or first-order spatial solitons are stable during propagation if they correspond to the stable branches of a power dispersion curve, as shown in the previous subsection. In practice, it is very difficult to launch an exact nonlinear mode or spatial soliton into an optical waveguide. Let us define a *quasi-mode* to be a field distribution that is close to an exact modal field distribution. Hence it is important to investigate the propagation behaviour of nonlinear quasi-modes as the mechanism of their propagation is much more complicated than that of their linear counterparts, where a simple interference algorithm applies.

We have performed propagation analyses of various nonlinear quasi-modes and generally have found that the deviation of the field profile during the propagation is comparable to the amount of the perturbation added to the "exact mode". A typical example is shown in Figure 7.18 for the nonlinear strip-loaded channel structure. The electric field distributions before and after propagation are given in Figure 7.19, where the initial field distribution corresponds to the upper branch of the solid power dispersion curve in Figure 7.1 at $\widetilde{P} = 60$ but with its actual power $\widetilde{P}$ scaled down to 44. As can be seen, the launch of such a quasi-nonlinear mode is not very critical. However, if the field pattern launched is far from a nonlinear mode, the nonlinear wave will become rather chaotic, justifying the usefulness of nonlinear "modal analysis".

7.4.4 All-optical switching

All-optical switching in nonlinear optical waveguides has been predicted theoretically, *e.g.* [251]. These analyses are restricted by the underlying assumption that the total field intensity distribution is a linear combination of individual modes which overlap, and do not correctly model strong nonlinearities. However, the field distribution has a great bearing on the matching of the optical switch to other optical devices connected to its input and output ports. Therefore, it is of practical importance to investigate the accurate propagation characteristics of field distributions in all-optical switching devices. Some work has already been reported for general 3D nonlinear waveguides [297].

Figure 7.20 illustrates all-optical switching from the nonlinear re-

$\bar{z}$

$\bar{y}$

$(\bar{x} = 0)$

-30   -12   0   15   30

Figure 7.18: The propagation of a nonlinear quasi-mode for 200 steps with $\Delta \bar{z} = 2$ and $\tilde{\beta} = 1.58$. Initial field distribution as for $\widetilde{P} = 60$ (upper solid curve of Figure 7.1) but with actual $\widetilde{P} = 44$.

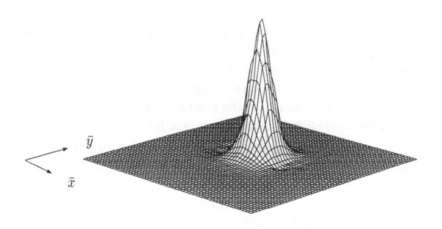

(a) Before propagation

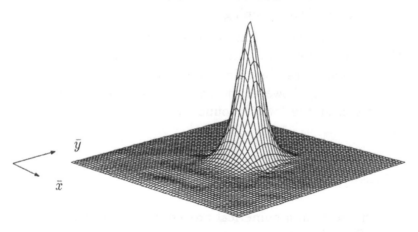

(b) After propagation

Figure 7.19: The scalar electric field distributions before and after propagation for the nonlinear quasi-mode whose propagation is shown in Figure 7.18.

gion into the linear guided region, where the initial field distribution corresponds to the upper portion of the solid power dispersion curve in Figure 7.1 at $\widetilde{P} = 34$ but with its actual power $\widetilde{P}$ scaled down to 32. Obviously, the field distribution is still well confined after the all-optical switching.

All-optical switching when the initial field is end-fired into the linear guided region is shown in Figure 7.21, where the initial field distribution corresponds to the lower portion of the solid power dispersion curve in Figure 7.1 at $\widetilde{P} = 54$ but with its actual power $\widetilde{P}$ scaled up to 56. As can be seen, the evolutional variation of the nonlinear wave is rather chaotic. Therefore, this configuration is not appropriate for all-optical switching.

In summary, for all-optical switching it is recommended that the initial field be launched in the nonlinear region with the field profile being approximately a nonlinear mode corresponding to the upper portion of the power dispersion curve. When the power is relatively high, the non-linear wave will remain in the nonlinear region; otherwise, it will switch to the linear guided region. An important feature of this arrangement is that the field distribution remains well confined during propagation before and after all-optical switching.

It should be mentioned that these results are somewhat dependent on the imposed "electric wall" boundary conditions. Therefore, further investigations with absorbing or infinite element boundaries is desirable. Qualitatively, however, no new phenomena are expected except that power near the "wall" boundary may radiate away causing power loss.

7.4.5   Conclusions

To conclude this section, a numerical procedure has been presented for investigating 3D nonlinear guided-wave propagation based on the scalar electric field approximation. A propagation stability analysis of non-linear modes has been performed for a nonlinear strip-loaded channel structure. As expected, those modes corresponding to the lower and the upper branches of the power dispersion curve are stable, whereas those associated with the middle branch are unstable. Thus, the bistability phenomenon predicted in § 6.4 is confirmed. The propagation behaviour of nonlinear quasi-modes corresponding to the stable branches of the power dispersion curve has also been investigated, showing that the variation of the field distribution during propagation is comparable to the

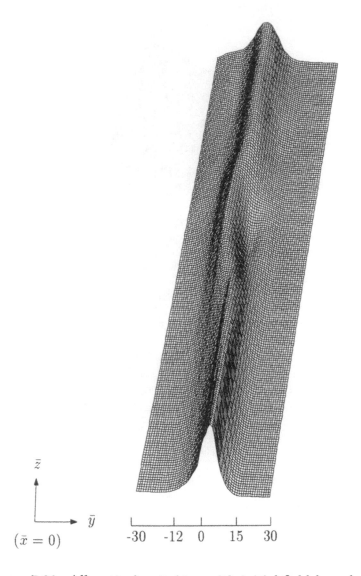

$\bar{z}$

$\bar{y}$

$(\bar{x} = 0)$

-30 -12 0 15 30

Figure 7.20: All-optical switching with initial field launched in the non-linear region (200 steps, $\Delta\bar{z} = 2$, $\tilde{\beta} = 1.54$, initial field distribution as for $\widetilde{P} = 34$ (upper solid curve of Figure 7.1) but with actual $\widetilde{P} = 32$).

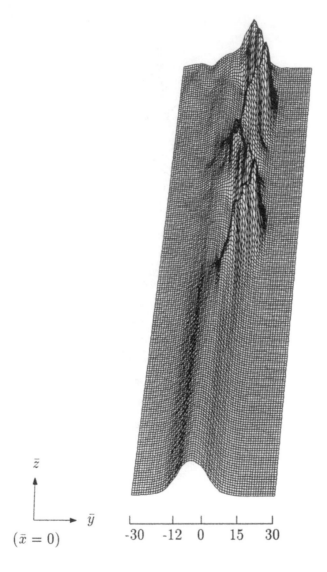

$\bar{z}$

$\bar{y}$

$(\bar{x} = 0)$

-30   -12   0   15   30

Figure 7.21: All-optical switching with initial field launched in the linear guided region (200 steps, $\Delta\bar{z} = 3$, $\tilde{\beta} = 1.56$, initial field distribution as for $\widetilde{P} = 54$ (lower solid curve of Figure 7.1) but with actual $\widetilde{P} = 56$).

amount of perturbation added to the "exact" mode. Consequently, nonlinear modal methods are justified and can be applied with confidence. In addition, in the context of all-optical switching, it has been demonstrated how to launch a field distribution which remains well confined during propagation.

Chapter 8

Coupled Optical Waveguides

8.1   Introduction

Linear and nonlinear coupled optical waveguides are most useful devices in optical signal processing [221], [306]-[309]. Linear optical directional couplers can be used as optical switches and power splitters. Nonlinear optical directional couplers have many potential applications in all-optical signal processing devices such as power-dependent switches, splitters and limiters, power filters, amplifiers, logic gates, *etc.*

In analysing coupled waveguides, three strategies are frequently used, namely, superposition of supermodes or arraymodes, coupled-mode theories and propagation methods. For quantitative characterisation of a coupler, the supermode superposition technique requires that both the propagation constants and modal fields of the composite waveguide are known *a priori*, whereas in coupled-mode theories it is assumed that the propagation constants and modal fields of each individual waveguide in isolation are known. In characterising a coupler with propagation methods, the field distribution input to one of the coupled waveguides is usually taken to be the fundamental modal field of that waveguide in isolation. Furthermore, for useful directional couplers, each waveguide in isolation should support only one or two guided modes. Therefore, all these three techniques cry out for modal analyses. However, optical waveguides whose modal fields can be solved analytically are rare. Thus, by arming the above three strategies with the numerical procedures presented in the previous chapters, a unique and powerful tool is created for analysing coupled optical waveguides.

Coupled optical waveguides may be configured in many ways. However, the most useful directional couplers normally consist of two single-mode waveguides placed close enough to achieve appreciable coupling effects. For simplicity, only these structures will be considered in the following. Nevertheless, all the three strategies to be introduced can be readily extended to cover coupled multiguide systems.

A related useful effect is mode coupling within an individual multimode (normally two-mode) waveguide. Such mode coupling is analogous to waveguide coupling and will not be treated separately.

The principle of directional couplers is described in §8.2. In §8.3 the supermode superposition technique is discussed. It is exact and efficient in computation but applicable to linear coupled waveguides only. It is the preferred technique for analysing linear waveguides with strong coupling. In §8.4 linear and nonlinear coupled-mode equations are derived by using a reciprocity approach. These equations are very general and may be applied to linear or nonlinear, isotropic or anisotropic, lossless or lossy, symmetric or asymmetric couplers provided that the waveguide materials are z-independent and non-magnetic. Note that coupled-mode theories using only individual guided modes as trial functions are not accurate when applied to coupled waveguides with strong coupling and/or strong nonlinear effects, and the coupled-mode theories including leaky radiating modes in the trial functions are normally intractable. Such coupled waveguides should be investigated by using propagation methods, which are discussed in §8.5. Propagation methods are inherently accurate, robust, and applicable to strongly coupled, strongly nonlinear and z-dependent couplers but are inefficient in computation. In §8.5 it is shown how to characterise linear and nonlinear coupled waveguides by propagation methods combined with modal methods. It is also shown how to improve the computational efficiency. An example of coupling between two identical cores is analysed in the linear case and in the nonlinear case. In addition, the application of FEM-FDM to spatial-temporal soliton systems is suggested.

8.2   Principle of Optical Couplers

For simplicity, we consider a planar directional coupler which is linear, lossless and z-independent. This coupler is formed by two identical planar waveguides, as shown in Figure 8.1. Furthermore, the coupler is assumed

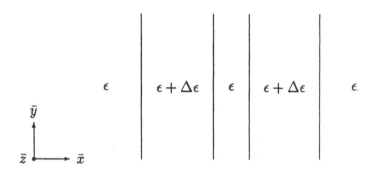

Figure 8.1: An example of a planar directional coupler.

to support only the two lowest TE supermodes shown in Figure 8.2. The first (lowest) supermode is called an even or symmetric mode and the second an odd or antisymmetric mode. Their electric modal fields are denoted by

$$E_y^+(\bar{x}, \bar{z}) = \psi^+(\bar{x})e^{-j\bar{\beta}^+\bar{z}} \tag{8.1}$$

and

$$E_y^-(\bar{x}, \bar{z}) = \psi^-(\bar{x})e^{-j\bar{\beta}^-\bar{z}} \tag{8.2}$$

for the even and odd modes, respectively, while $\bar{\beta}^+$ and $\bar{\beta}^-$ are the propagation constants normalised with respect to $\rho = k_0$ as in Chapter 7. For lossless media $\psi^+(\bar{x})$ and $\psi^-(\bar{x})$ are real functions, and so are the normalised propagation constants, and $\bar{\beta}^+ \geq \bar{\beta}^- > 0$. The amplitudes of the modal fields are normalised such that

$$\int_{-\infty}^{+\infty} |\psi^+(\bar{x})|^2 d\bar{x} = \int_{-\infty}^{+\infty} |\psi^-(\bar{x})|^2 d\bar{x} = 1. \tag{8.3}$$

The total electric field in the coupler is given by

$$\begin{aligned} E_y(\bar{x}, \bar{z}) &= a^+ E_y^+(\bar{x}, \bar{z}) + a^- E_y^-(\bar{x}, \bar{z}) \\ &= a^+ \psi^+(\bar{x})e^{-j\bar{\beta}^+\bar{z}} + a^- \psi^-(\bar{x})e^{-j\bar{\beta}^-\bar{z}} \end{aligned} \tag{8.4}$$

where a^+ and a^- are arbitrary constants but are assumed to be real without loss of generality.

As $\bar{\beta}^+ \neq \bar{\beta}^-$ for a practical coupler, there exists a point on the z-axis where the even and odd modes are reinforcing in Guide 1 and are in

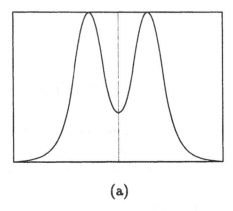

(a)

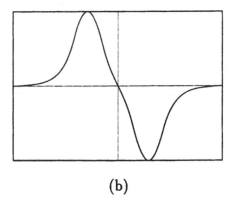

(b)

Figure 8.2: The modal electric fields of the two lowest supermodes of a symmetric planar coupler: (a) the even mode and (b) the odd mode.

opposition in Guide 2. Furthermore, if $a^+ = a^-$, almost all of the optical power appears in Guide 1. After propagating a (normalised) distance of

$$\bar{L}_c = \frac{\pi}{\bar{\beta}^+ - \bar{\beta}^-}, \qquad (8.5)$$

these two modes are reinforcing in Guide 2 and are in opposition in Guide 1. Then, almost all of the optical power appears in Guide 2. After propagating a further (normalised) distance of $\bar{L}_c$, the optical power reappears in Guide 1. The principle of a directional coupler is based on this phenomenon of beating or interference between the two (or more for multiple waveguide coupling) supermodes. The (normalised) distance $\bar{L}_c$ is called the (normalised) *coupling length* or (normalised) *beat length* as it equals the distance over which the guided power transfers from one of the individual guides into the other. The beat length is the most important parameter to characterise a directional coupler.

When an optical beam is incident on, for example, Guide 1, it will excite both the even and the odd supermodes as well as leaky radiating modes. By using the modal expansion method together with mode orthogonality, the power coupled to each supermode can be uniquely determined [8, 310]. The leaky radiating modes will die away after a certain distance of propagation as we have assumed that the composite waveguide supports only the two lowest guided supermodes. The power will switch to Guide 2 after an odd number of beat lengths of propagation (*cross state*), and back to Guide 1 after an even number of beat lengths of propagation (*bar state*).

It should be mentioned that complete power transfer from one individual guide to the other is strictly impossible. This is because the even and odd supermodes cannot completely cancel each other in an individual guide because of the difference between their field distributions within a single guide. As a rule of thumb, the stronger the coupling, the shorter the beat length, and the greater the difference between the modal field distributions of the two supermodes within a single guide, and thus the worse the degree of switching.

The above describes the fundamental principles of linear directional couplers. The operation of nonlinear directional couplers is more complicated. Particularly, the beat length and field distributions of the composite structure are power dependent. Thus it is possible to use one beam to control another for all-optical signal processing.

8.3   Supermode Superposition Technique

The supermode or arraymode superposition technique is perhaps the most accurate and also the simplest method to characterise linear coupled parallel waveguides. Its principle has been fully described in the previous section, where for the convenience of description we confined ourselves to planar couplers supporting only two TE modes. However, the supermode superposition technique itself can equally be applied to any linear coupled parallel waveguides supporting any kind of guided modes. The mathematical description of the interference between supermodes is straightforward. The following is mainly concerned with the computational aspects.

For linear coupled waveguides consisting of two parallel guiding regions, the computation of the two lowest modes of the composite structure is simple and similar to that given in §5.5. In particular, when a coupler is formed by two identical waveguides, use can be made of the symmetry property by solving only half the structure. Figure 8.3 illustrates an example of symmetric coupled waveguides and their symmetry plane. When solving half the structure, the symmetry plane becomes a boundary. The boundary condition at the symmetry plane behaves like a magnetic wall[1] for the even mode and like an electric wall[2] for the odd mode. For the even mode only the normal component of the electric flux density D (or the electric field E in isotropic media) in the E formulation or the tangential components of the magnetic field H in the H formulation need to be set to zero at the symmetry plane when using the intermediate form of the weak formulation, and the other boundary conditions are treated as natural boundary conditions. Similarly, for the odd mode only the normal component of the magnetic flux density B (or the magnetic field H in isotropic media) in the H formulation or the tangential components of the electric field E in the E formulation need to be set to zero. With the weak-guidance approximation using the scalar E-field formulation, the symmetry plane is treated as a natural boundary condition for the even mode and as an essential (homogeneous Dirichlet) boundary condition for the odd mode when using the intermediate form of the weak formulation.

---

[1] The normal component of the electric flux density and the tangential components of the magnetic field vanish, that is, $n \cdot D = 0$ and $n \times H = 0$.

[2] The tangential components of the electric field and the normal component of the magnetic flux density vanish, that is, $n \times E = 0$ and $n \cdot B = 0$.

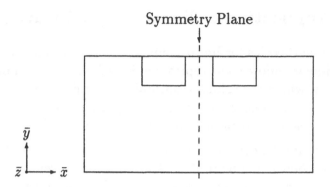

Figure 8.3: An example of symmetric coupled waveguides and their symmetry plane.

It is known that nonlinear modes are not subject to superposition. Thus, the prescribed supermode superposition technique, though the most accurate technique for analysing linear couplers, cannot be directly applied to the analysis of nonlinear couplers.

8.4 Coupled-Mode Theory

The theory of mode coupling was first developed by Pierce [311] to analyse the coupling of electron beam waves and waves on electromagnetic structures in electron beam tubes [312]. Later, coupled-mode theory was applied to optical waveguides [313]-[316]. In the past ten years application of coupled-mode theory to linear and nonlinear optical waveguides has attracted world-wide interest, *e.g.* [245], [317]-[332]. The aim of coupled-mode theory is to transform partial-differential wave equations into a set of coupled ordinary differential equations, where the complex field amplitudes (functions of the longitudinal coordinate) in the individual waveguides are to be found. Coupled-mode equations may be derived from a reciprocity approach [323], a variational approach [312], power conservation [252], or simply substituting the trial field into the Maxwell equations [321, 322]. The resulting coupled-mode equations are virtually the same. The differences lie in the coupling coefficients.

8.4.1   Application to linear coupled waveguides

Coupled-mode theories for linear coupled waveguides have been well developed. Here we quote some typical published results as an introduction to the theory, which are valid for isotropic waveguides. The derivation of linear and then nonlinear coupled-mode equations, valid for anisotropic and lossy waveguides, is presented in the next subsection.

As an example, consider two parallel cores embedded in a common cladding, as shown in Figure 8.4. Assume each waveguide in isolation (the other waveguide having been replaced by cladding) supports only one guided mode, denoted by

$$[\boldsymbol{E}^{(p)}(\bar{x},\bar{y}), \boldsymbol{H}^{(p)}(\bar{x},\bar{y})]\exp(-j\bar{\beta}^{(p)}\bar{z}) \qquad \forall p \in \{1,2\} \qquad (8.6)$$

Each of these waveguides is characterised by a relative permittivity profile $\epsilon_r^{(p)}(\bar{x},\bar{y})$. The whole system is characterised by a composite relative permittivity profile $\epsilon_r(\bar{x},\bar{y})$ which can be expressed as

$$\epsilon_r(\bar{x},\bar{y}) = \epsilon_r^{(p)}(\bar{x},\bar{y}) + \Delta\epsilon_r^{(p)}(\bar{x},\bar{y}) \qquad \forall p \in \{1,2\} \qquad (8.7)$$

where $\Delta\epsilon_r^{(p)}$ corresponds to the perturbation of waveguide (p) due to the other core.

The basic assumption in coupled-mode theory is that the field in the whole structure can be written as a mere linear combination of unperturbed modes, that is,

$$[\boldsymbol{E}, \boldsymbol{H}] = \sum_{p=1}^{2} a_p(\bar{z})[\boldsymbol{E}^{(p)}(\bar{x},\bar{y}), \boldsymbol{H}^{(p)}(\bar{x},\bar{y})]. \qquad (8.8)$$

One may also expand only the transverse field components by writing

$$[\boldsymbol{E}_T, \boldsymbol{H}_T] = \sum_{p=1}^{2} a_p(\bar{z})[\boldsymbol{E}_T^{(p)}(\bar{x},\bar{y}), \boldsymbol{H}_T^{(p)}(\bar{x},\bar{y})]. \qquad (8.9)$$

The corresponding longitudinal components are obtained by substituting the above equation into the appropriate Maxwell equations [325], yielding

$$[\boldsymbol{E}_z, \boldsymbol{H}_z] = \sum_{p=1}^{2} a_p(\bar{z})[\frac{\epsilon_r^{(p)}}{\epsilon_r}\boldsymbol{E}_z^{(p)}(\bar{x},\bar{y}), \boldsymbol{H}_z^{(p)}(\bar{x},\bar{y})]. \qquad (8.10)$$

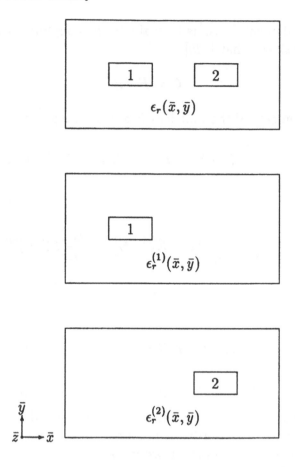

Figure 8.4: An example of two parallel coupled waveguides.

Then a set of linear coupled-mode equations of first-order approximation are assumed for the modal amplitudes:

$$\frac{da_p}{d\bar{z}} = -j\bar{\beta}^{(p)} a_p - j\sum_{q=1}^{2} C_{pq}\, a_q \qquad \forall p \in \{1,2\} \qquad (8.11)$$

or in matrix form:

$$\frac{d\underline{a}}{d\bar{z}} = -j(B+C)\,\underline{a} \qquad (8.12)$$

where

$$\underline{a} = [a_1, a_2]^T, \qquad (8.13)$$

and

$$B = \begin{bmatrix} \bar{\beta}_1 & 0 \\ 0 & \bar{\beta}_2 \end{bmatrix}, \qquad C = \begin{bmatrix} C_{11} & C_{12} \\ C_{21} & C_{22} \end{bmatrix}. \qquad (8.14)$$

The fundamental task is to evaluate the coupling coefficient matrix C. It can be shown that [325]

$$C = P^{-1}K, \tag{8.15}$$

where the elements of the matrices P and K are given by

$$P_{pq} = \frac{1}{4}\int_{\Omega}(\boldsymbol{E}_T^{(p)} \times \boldsymbol{H}_T^{(q)} - \boldsymbol{H}_T^{(p)} \times \boldsymbol{E}_T^{(q)}) \cdot \boldsymbol{a}_z \, d\bar{x} \, d\bar{y} \tag{8.16}$$

and

$$K_{pq} = \frac{1}{4}\omega\epsilon_0\int_{\Omega}\Delta\epsilon_r^{(q)}(\boldsymbol{E}_T^{(p)} \cdot \boldsymbol{E}_T^{(q)} - \frac{\epsilon_r^{(p)}}{\epsilon_r}E_z^{(p)}E_z^{(q)}) \, d\bar{x} \, d\bar{y} \tag{8.17}$$

when using the transverse field expansion for a trial field (Equations 8.9 and 8.10); or

$$K_{pq} = \frac{1}{4}\omega\epsilon_0\int_{\Omega}\Delta\epsilon_r^{(q)}(\boldsymbol{E}_T^{(p)} \cdot \boldsymbol{E}_T^{(q)} - E_z^{(p)}E_z^{(q)}) \, d\bar{x} \, d\bar{y} \tag{8.18}$$

when using the full vector field expansion for a trial field (Equation 8.8).

Note that the above coupling coefficients are derived from vector modal fields with a first-order approximation. Coupled-mode equations may also be derived from higher-order approximations [327] and/or from scalar modal fields under the weak-guidance approximation [322, 329] to suit individual needs.

The solution of linear coupled-mode equations is straightforward as long as the propagation constants of the individual isolated waveguides and all the coupling coefficients are known [329]. In turn the coupling coefficients require the modal fields of the individual waveguides. Since optical waveguides whose propagation constants and modal fields can be solved analytically are rare, the numerical procedures presented in the previous chapters should be very useful for finding such modal fields and propagation constants.

For accurate analysis of coupled waveguides, three precautionary factors should be considered when applying coupled-mode theories:

- When the individual waveguides forming the coupler are not weakly-guiding, polarisation birefringence must be considered, and thus vector coupled-mode theories are mandatory for such structures.

- When the coupling is relatively strong, the fundamental modes of the two individual waveguides are not power orthogonal, and thus nonorthogonal coupled-mode formulations are required.

- When the coupling is sufficiently strong, the supermodes of a composite waveguide cannot be represented accurately by only the fundamental modes of the individual waveguides in isolation, and thus all guided and leaky radiating modes should be included, which renders the coupled-mode theory intractable; propagation methods are then much preferred.

8.4.2   Application to nonlinear coupled waveguides

The power-dependent property of nonlinear coupled optical waveguides has potential applications in ultra-fast all-optical signal processing. Coherent nonlinear directional couplers are good candidates for all-optical switches, which are essential for all-optical networks. To simulate weak nonlinear effects in coupled waveguides, nonlinear coupled-mode theory can be applied. Following the pioneering work of Jensen in 1982 [245], nonlinear coupled-mode theory has been further developed by many researchers for analysing nonlinear couplers, *e.g.* [245]-[258], [331, 333, 334]. The nonlinear coupled-mode equations reported in the literature are generally derived from scalar waves or TE modes for coupling between identical waveguides. They are useful for weakly-guiding and/or planar couplers. In some early works certain nonlinear terms were ignored, leading to inconsistency in the theory and reduced accuracy of the predicted results, as pointed out in [331, 334]. In the following, nonlinear coupled-mode equations are derived from a conjugated reciprocity approach. They are valid for nonlinear directional couplers where the coupled modes are not necessarily TE modes and the coupled waveguides are not necessarily identical.

Consider two sets of solutions (E_1, H_1) and (E_2, H_2) of the Maxwell equations:

$$\nabla \times E = -j\omega\mu_0 H \qquad (8.19)$$

$$\nabla \times H = j\omega\epsilon_0\hat{\epsilon}_r \cdot E \qquad (8.20)$$

with the associated interface and boundary conditions, where the relative permittivity $\hat{\epsilon}_r$ is a function of the electric field E and the medium is different for each set. Also, the common phase factor $\exp(j\omega t)$ is implied for both sets. We define a vector function $\boldsymbol{F}$ in a conjugated form by

$$\boldsymbol{F} \overset{\triangle}{=} \boldsymbol{E}_1^* \times \boldsymbol{H}_2 - \boldsymbol{H}_1^* \times \boldsymbol{E}_2. \tag{8.21}$$

The two-dimensional divergence theorem of Gauss states that [8]

$$\frac{d}{d\bar{z}} \int_\Omega \boldsymbol{F} \cdot \boldsymbol{a}_z \, d\bar{x} \, d\bar{y} = \int_\Omega \overline{\nabla} \cdot \boldsymbol{F} \, d\bar{x} \, d\bar{y} - \oint_C \boldsymbol{F} \cdot \boldsymbol{n} \, d\bar{l}, \tag{8.22}$$

where C is the boundary of the cross-section Ω and $\boldsymbol{n}$ is the outward unit vector normal to C.

Using the vector identity

$$\nabla \cdot (\boldsymbol{A} \times \boldsymbol{B}) = \boldsymbol{B} \cdot (\nabla \times \boldsymbol{A}) - \boldsymbol{A} \cdot (\nabla \times \boldsymbol{B}), \tag{8.23}$$

we obtain

$$\overline{\nabla} \cdot \boldsymbol{F} = j\omega\epsilon_0 \boldsymbol{E}_1^* \cdot (\hat{\epsilon}_{r1}^\dagger - \hat{\epsilon}_{r2}) \cdot \boldsymbol{E}_2. \tag{8.24}$$

In the following the line integral term in Equation 8.22 is assumed to be negligible for bounded modes by a choosing sufficiently large Ω. Then, substituting Equations 8.19 to 8.21 and 8.24 into Equation 8.22 yields

$$\frac{d}{d\bar{z}} \int_\Omega (\boldsymbol{E}_{T1}^* \times \boldsymbol{H}_{T2} - \boldsymbol{H}_{T1}^* \times \boldsymbol{E}_{T2}) \cdot \boldsymbol{a}_z \, d\bar{x} \, d\bar{y}$$

$$= j\omega\epsilon_0 \int_\Omega \boldsymbol{E}_1^* \cdot (\hat{\epsilon}_{r1}^\dagger - \hat{\epsilon}_{r2}) \cdot \boldsymbol{E}_2 \, d\bar{x} \, d\bar{y}, \tag{8.25}$$

where the subscript T denotes the transverse components. This reciprocity relation is applicable to any pair of dielectric media and is exact provided that $(\boldsymbol{E}_1, \boldsymbol{H}_1)$ and $(\boldsymbol{E}_2, \boldsymbol{H}_2)$ satisfy Equations 8.19 and 8.20 as well as the associated discontinuity and boundary conditions.

Next, consider a nonlinear directional coupler formed by two parallel waveguides denoted by (u) and (v), which are not necessarily identical. Assume that each waveguide in isolation supports only one guided mode: $(\boldsymbol{E}^{(u)}, \boldsymbol{H}^{(u)})$ for waveguide (u) and $(\boldsymbol{E}^{(v)}, \boldsymbol{H}^{(v)})$ for waveguide (v).

The first set of solutions is chosen to be the guided mode in the isolated waveguide (u), namely

$$\boldsymbol{E}_1 = \boldsymbol{E}^{(u)} \exp\{-j\bar{\beta}^{(u)}\bar{z}\} \tag{8.26}$$

$$\boldsymbol{H}_1 = \boldsymbol{H}^{(u)} \exp\{-j\bar{\beta}^{(u)}\bar{z}\} \tag{8.27}$$

so that

$$\hat{\epsilon}_{r1}(\bar{x}, \bar{y}) = \hat{\epsilon}_r^{(u)}(\bar{x}, \bar{y}). \tag{8.28}$$

The second set of solutions is chosen to be the combination

$$\boldsymbol{E}_2 = a_u(\bar{z})\boldsymbol{E}^{(u)}\exp\{-j\bar{\beta}^{(u)}\bar{z}\} + a_v(\bar{z})\boldsymbol{E}^{(v)}\exp\{-j\bar{\beta}^{(v)}\bar{z}\} \tag{8.29}$$
$$\boldsymbol{H}_2 = a_u(\bar{z})\boldsymbol{H}^{(u)}\exp\{-j\bar{\beta}^{(u)}\bar{z}\} + a_v(\bar{z})\boldsymbol{H}^{(v)}\exp\{-j\bar{\beta}^{(v)}\bar{z}\} \tag{8.30}$$

where the relative permittivity dyadic is taken as

$$\hat{\epsilon}_{r2}(\bar{x}, \bar{y}) = \hat{\epsilon}_r(\bar{x}, \bar{y}) \tag{8.31}$$

and $\hat{\epsilon}_r(\bar{x}, \bar{y})$ is the relative permittivity dyadic of the composite waveguide. These relations are just the modal expansions in terms of the full-vector guided modes in waveguides (u) and (v). The expansion is only an approximate solution of the composite waveguide with the leaky radiating modes being neglected.

Substituting Equations 8.26 to 8.31 into Equation 8.25 yields one of the coupled-mode equations:

$$jP_{uu}\frac{da_u}{d\bar{z}} + jP_{uv}\exp\{j(\bar{\beta}^{(u)} - \bar{\beta}^{(v)})\bar{z}\}\frac{da_v}{d\bar{z}} = (K_{uu} + \Delta\bar{\beta}_{uu}P_{uu})\,a_u$$
$$+ (K_{uv} + \Delta\bar{\beta}_{uv}P_{uv})\exp\{j(\bar{\beta}^{(u)} - \bar{\beta}^{(v)})\bar{z}\}\,a_v. \tag{8.32}$$

Similarly, using the guided mode in the isolated waveguide (v) as the first set of solutions yields the other coupled-mode equation:

$$jP_{vu}\exp\{j(\bar{\beta}^{(v)} - \bar{\beta}^{(u)})\bar{z}\}\frac{da_u}{d\bar{z}} + jP_{vv}\frac{da_v}{d\bar{z}} = (K_{vu} + \Delta\bar{\beta}_{vu}P_{vu})$$
$$\exp\{j(\bar{\beta}^{(v)} - \bar{\beta}^{(u)})\bar{z}\}\,a_u + (K_{vv} + \Delta\bar{\beta}_{vv}P_{vv})\,a_v \tag{8.33}$$

where

$$P_{pq} \triangleq \frac{1}{4}\int_\Omega (\boldsymbol{E}_T^{(p)\,*} \times \boldsymbol{H}_T^{(q)} - \boldsymbol{H}_T^{(p)\,*} \times \boldsymbol{E}_T^{(q)}) \cdot \boldsymbol{a}_z\, d\bar{x}\, d\bar{y}, \tag{8.34}$$

$$K_{pq} \triangleq -\frac{1}{4}\omega\epsilon_0\int_\Omega \boldsymbol{E}^{(p)\,*} \cdot (\hat{\epsilon}_r^{(p)\,\dagger} - \hat{\epsilon}_r) \cdot \boldsymbol{E}^{(q)}\, d\bar{x}\, d\bar{y}, \tag{8.35}$$

and

$$\Delta\bar{\beta}_{pq} \triangleq \bar{\beta}^{(p)\,*} - \bar{\beta}^{(q)}, \qquad \forall p, q \in \{u, v\}. \tag{8.36}$$

These coupled-mode equations are very general. They are valid for linear or nonlinear, isotropic or anisotropic, lossless or lossy, symmetric or asymmetric waveguides as long as the media involved are z-independent and non-magnetic. Following similar procedures given in

[258, 312, 320, 330], these equations can be readily extended to multi-waveguide systems. For z-dependent coupled waveguides coupled local-mode theory should be employed [8]. For nonlinear coupled waveguides the nonlinearity of $\hat{\epsilon}$ should be considered in evaluating K_{pq}, a matter to be discussed shortly.

When $\bar{\beta}^{(u)}$ and $\bar{\beta}^{(v)}$ are not identical, the coupling coefficients will be z-dependent, which is in agreement with the variational coupled-mode theory [329]. Note that when the media involved are lossy,

$$\Delta\bar{\beta}_{pq} \neq 0, \qquad\qquad p, q \in \{u, v\} \tag{8.37}$$

even if the two coupled waveguides are identical.

Up to this point the only approximation that has been made is in writing Equations 8.29 and 8.30. When the two coupled waveguides are identical and the media are lossless, these coupled-mode equations reduce to the following familiar form, *e.g.* [312]:

$$j \left[\begin{array}{cc} P_{uu} & P_{uv} \\ P_{vu} & P_{vv} \end{array} \right] \left[\begin{array}{c} da_u/d\bar{z} \\ da_v/d\bar{z} \end{array} \right] = \left[\begin{array}{cc} K_{uu} & K_{uv} \\ K_{vu} & K_{vv} \end{array} \right] \left[\begin{array}{c} a_u \\ a_v \end{array} \right]. \tag{8.38}$$

Note that the coupled-mode equations will be slightly different from the above if the phase factor $\exp\{-j\bar{\beta}\bar{z}\}$ is incorporated in the complex amplitudes $a_u(\bar{z})$ and $a_v(\bar{z})$.

Though Equations 8.32 and 8.33 are applicable to nonlinear coupled waveguides, it is customary to express the nonlinearity explicitly by simply expanding the coupling coefficients K_{pq} in terms of the complex amplitudes $a_u(\bar{z})$ and $a_v(\bar{z})$. For Kerr nonlinearity the nonlinear relative permittivities $\hat{\epsilon}_r$, $\hat{\epsilon}_r^{(u)}$ and $\hat{\epsilon}_r^{(v)}$ in Equation 8.35 may be expressed as a linear term plus an isotropic nonlinear perturbation term:

$$\hat{\epsilon}_r^{(u)} = \hat{\epsilon}_r^{l\,(u)} + \alpha^{(u)} \|\boldsymbol{E}^{(u)}\|^2 \hat{I} \tag{8.39}$$

$$\hat{\epsilon}_r^{(v)} = \hat{\epsilon}_r^{l\,(v)} + \alpha^{(v)} \|\boldsymbol{E}^{(v)}\|^2 \hat{I} \tag{8.40}$$

$$\tag{8.41}$$

and

$$\hat{\epsilon}_r = \hat{\epsilon}_r^{l} + \alpha \|a_u \boldsymbol{E}^{(u)} + a_v \boldsymbol{E}^{(v)}\|^2 \hat{I} \tag{8.42}$$

where $\hat{I}$ is the identity dyadic and the squared norm is defined as

$$\|\boldsymbol{v}\|^2 \overset{\triangle}{=} \boldsymbol{v}^* \cdot \boldsymbol{v} \tag{8.43}$$

and the nonlinear coefficients α, $\alpha^{(u)}$ and $\alpha^{(v)}$ are related to the usual nonlinear optical coefficient n_2 by Equation 6.3[3]. Here the modal fields

---

[3]The symbol α is used here in place of a in Equation 6.3.

$\boldsymbol{E}^{(u)}$ and $\boldsymbol{E}^{(v)}$ may be interpreted as linear modes scaled by using appropriate powers or as nonlinear modes where $\boldsymbol{E}^{(u)}(\bar{x}, \bar{y})$ and $\boldsymbol{E}^{(v)}(\bar{x}, \bar{y})$ are power dependent. Improved accuracy is expected by using nonlinear modes, which has been investigated and reported in [331]. However, the power conservation relation $|a_u(\bar{z})|^2 + |a_v(\bar{z})|^2 = 1$ used in [331] is valid only for the coupling of power-orthogonal modes or the weak coupling of the modes from two individual waveguides where the cross-power terms can be ignored. Strictly speaking, the fundamental modes of two individual waveguides can never be power-orthogonal as long as there is any coupling between the two modes. The correct power conservation relation reads

$$
\begin{aligned}
P_c &= \frac{1}{2k_0^2} \cdot \Re\left\{ \int_\Omega (\boldsymbol{E}_2 \times \boldsymbol{H}_2^*) \cdot \boldsymbol{a}_z \, d\bar{x} \, d\bar{y} \right\} \\
&= \Re\left\{ \sum_{p \in \{u,v\}} \sum_{q \in \{u,v\}} S_{pq} \exp\{-j(\bar{\beta}^{(p)} - \bar{\beta}^{(q)\,*})\bar{z}\} \, a_p a_q^* \right\} \quad (8.44)
\end{aligned}
$$

where P_c is the total power guided in the composite waveguide, and $\Re\{\cdot\}$ denotes the real part of its argument, and

$$
S_{pq} = \frac{1}{2k_0^2} \int_\Omega (\boldsymbol{E}_T^{(p)} \times \boldsymbol{H}_T^{(q)\,*}) \cdot \boldsymbol{a}_z \, d\bar{x} \, d\bar{y} \qquad \forall p, q \in \{u, v\}. \quad (8.45)
$$

Note that the cross-power terms $S_{pq} \neq 0$ $(p \neq q)$ in general.

Substituting Equations 8.39 to 8.42 into Equation 8.35 yields

$$
K_{pq} = K_{pq}^l + \alpha \sum_{r \in \{u,v\}} \sum_{s \in \{u,v\}} N_{pq}^{rs} a_r^* a_s - \alpha^{(p)\,*} N_{pq}^{pp}
$$
$$
\forall p, q \in \{u, v\} \quad (8.46)
$$

where

$$
K_{pq}^l \triangleq -\frac{1}{4}\omega\epsilon_0 \int_\Omega \boldsymbol{E}^{(p)\,*} \cdot (\hat{\epsilon}_r^{l\,(p)\,\dagger} - \hat{\epsilon}_r^l) \cdot \boldsymbol{E}^{(q)} \, d\bar{x} \, d\bar{y} \quad (8.47)
$$

$$
N_{pq}^{rs} \triangleq \frac{1}{4}\omega\epsilon_0 \int_\Omega (\boldsymbol{E}^{(r)\,*} \cdot \boldsymbol{E}^{(s)})(\boldsymbol{E}^{(p)\,*} \cdot \boldsymbol{E}^{(q)}) \, d\bar{x} \, d\bar{y}. \quad (8.48)
$$

The last term in Equation 8.46 would have disappeared if we had used the linear solutions of the two waveguides as the first set of solutions.

The explicit nonlinear coupled-mode equations are obtained by substituting Equations 8.46 to 8.48 into Equations 8.32 and 8.33 or into

Equation 8.38 for lossless and symmetrical couplers. They can be written formally as

$$j \sum_{q \in \{u,v\}} P_{pq} \exp\{j(\beta^{(p)} - \beta^{(q)})\bar{z}\} \frac{da_q}{d\bar{z}} = \sum_{q \in \{u,v\}} [K_{pq}^l + \Delta\beta_{pq} P_{pq}$$

$$+ \alpha \sum_{r \in \{u,v\}} \sum_{s \in \{u,v\}} N_{pq}^{rs} a_r^* a_s - \alpha^{(p)*} N_{pq}^{pp}] \exp\{j(\beta^{(p)} - \beta^{(q)})\bar{z}\} a_q$$

$$p \in \{u,v\} \qquad (8.49)$$

for general nonlinear coupled waveguides, or

$$j \sum_{q \in \{u,v\}} P_{pq} \frac{da_q}{d\bar{z}} = \sum_{q \in \{u,v\}} [K_{pq}^l + \alpha \sum_{r \in \{u,v\}} \sum_{s \in \{u,v\}} N_{pq}^{rs} a_r^* a_s - \alpha^{(p)} N_{pq}^{pp}] a_q$$

$$p \in \{u,v\} \qquad (8.50)$$

for lossless and symmetric nonlinear coupled waveguides. These coupled-mode equations are consistent with those reported in [335] for isotropic and lossless coupled waveguides.

Nonlinear coupled-mode equations may be solved analytically by using Stokes parameters and Jacobian elliptic functions [245, 254, 331] or simply by numerical integration [246, 256, 333, 334]. With the analytical approach the power conservation relation (Equation 8.44) is incorporated, while with the numerical integration appropriate initial values $a_u(0)$ and $a_v(0)$ are chosen according to the total guided power (Equation 8.44). The coupling coefficients can be evaluated by the modal analysis presented in Chapters 5 and 6 together with numerical integration.

Nonlinear coupled-mode theories are useful tools for characterising nonlinear coupled waveguides and are also efficient in computation. However, caution should be exercised when applying them to coupled waveguides with strong nonlinear effects. Specifically, they are inaccurate for the coupling of those nonlinear modes corresponding to the upper portion of the power dispersion curve (*see* Chapter 6) unless the trial fields used in deriving them include both guided and leaky radiating modes, which is intractable. Coupled waveguides with strong nonlinear effects should be investigated by propagation methods. In addition, the precautionary factors outlined for linear coupled waveguides also apply here.

In this section the discussion has been restricted to z-independent and non-dispersive coupled waveguides. The methodology described can be readily extended to cover corrugated waveguides and couplers [336]-[338] as well as directional couplers with dispersive susceptibilities [339].

8.5   Propagation Method

Propagation methods have been well developed for analysing linear and nonlinear optical waveguides [260]-[290], [340]-[346]. They may be classified as the *beam propagation method* (BPM), which is also known as the *split-step Fourier method*, the *finite-difference propagation method* (FDPM) and the *finite-element method plus finite-difference method* (FEM-FDM). Advantages and disadvantages of each method are discussed in [271].

Unlike the supermode superposition technique and standard coupled-mode theories, which are applicable to z-independent waveguides only, propagation methods can be equally applied to z-dependent waveguides. Application of propagation methods to coupled waveguides is virtually the same as to individual waveguides. Note that propagation methods are generally intensive in computation compared with the other two approaches.

Coupled waveguides, regardless of strong coupling or strong nonlinear effects, can be fully and accurately characterised by propagation methods, whereby both the beat length and optical power distribution along the coupler can be found by examining the evolution of the field distribution. Unlike in coupled-mode theories where the trial fields have to be the guided modes of the isolated waveguides, there is no restriction on the initial field distribution launched into one of the coupled waveguides when using propagation methods. However, in order to characterise a coupler, it is preferred that the initial field distribution closely match the guided mode of the isolated waveguide into which the field is to be launched. This is particularly important for waveguides with strong coupling and/or strong nonlinear effects, especially when using a reflecting boundary. Otherwise, the field evolution can be rather chaotic, and it is difficult to find the beat length.

8.5.1   Computational aspects

The propagation algorithms and procedures described in §7.4 can be directly applied to linear and nonlinear coupled waveguides. For example, consider the coupled waveguide structure shown in Figure 8.4, and assume that a field distribution will be launched in Guide 1. Then the propagation analysis can be performed in the following way:

(1) Set the permittivity of Guide 2 to that of the cladding and compute the guided mode of Guide 1. Here the appropriate guided power should be incorporated into the computation if the coupler is nonlinear.

(2) Restore the permittivity of Guide 2.

(3) Perform the propagation analysis using the modal field distribution computed in Step (1) as the initial field and the associated propagation constant as the reference propagation constant.

(4) Find the beat length and evaluate the power distribution between the two coupled waveguides along the propagation direction by examining the field evolution.

The use of the propagation constant of the initial modal field as the reference propagation constant is vital for the efficiency of computation. For example, the propagation step can be chosen as large as thousands of wavelengths for weakly coupled waveguides with weak nonlinear effects as long as the beat length is much larger than this value. On the other hand, if the initial field is far from a guided mode and/or the reference propagation constant is far from that of the initial modal field, the complex field amplitude will change rapidly during propagation, and the propagation step must be kept small. Also, when the initial field is far from a guided mode, considerable energy will be radiated away, and therefore open boundary conditions must be imposed [26]-[36], [347]-[349].

8.5.2   A numerical example

As an example, consider the coupled waveguide structure shown in Figure 8.4 and use the weak-guidance approximation to perform the analysis. Using the normalised coordinates and electric field, the waveguide parameters are chosen as follows. Two identical square cores are 20×20 in size and have a linear relative permittivity of 1.53^2. The centre-to-centre separation of the two cores is 36. The cladding is 120×90 in size and has a nonlinear relative permittivity of $1.52^2 + |\widetilde{E}|^2/(1 + |\widetilde{E}|^2)$. The coupled waveguide structure is meshed into 1044 first-order elements with 565 nodes, as shown in Figure 8.5.

In the linear case, this composite waveguide supports the two supermodes shown in Figure 8.6. The coupling between the two individual waveguides results in the beating between these two supermodes. In

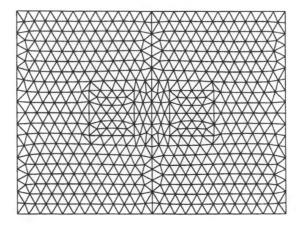

Figure 8.5: Mesh network of the coupled waveguide structure (1044 first-order elements, 565 nodes).

order to perform the propagation analysis, the guided mode of Guide 1 in isolation is computed by using the modal analysis, and the result is shown in Figure 8.7. This modal field distribution and its propagation constant are chosen as the initial field and reference propagation constant, respectively, for the propagation analysis.

The following propagation results are plots of the electric field intensity squared and integrated over the vertical- or y-domain, that is,

$$F(\bar{x}, \bar{z}) = \int_{-45}^{45} |\tilde{E}(\bar{x}, \bar{y}, \bar{z})|^2 \, d\bar{y}. \tag{8.51}$$

The guided powers of the individual waveguides are defined by splitting the total guided power at the vertical symmetry plane.

Figure 8.8 shows the field evolution along the composite waveguide for 200 steps with a step size $\Delta\bar{z} = 100.0$. The periodic power switching is clearly shown in Figure 8.9. The coupling length is found to be about $45\Delta\bar{z}$, subject to the discretization error of meshing the structure.

In the nonlinear case, the normalised guided power $\tilde{P}$ is chosen to be 30.0 without loss of generality. The nonlinear modal field of Guide 1 is very similar to the linear one shown in Figure 8.7 (only slightly "fatter"). This is because the core itself is linear and the nonlinear action is still weak. The field evolution and periodic power switching are shown in

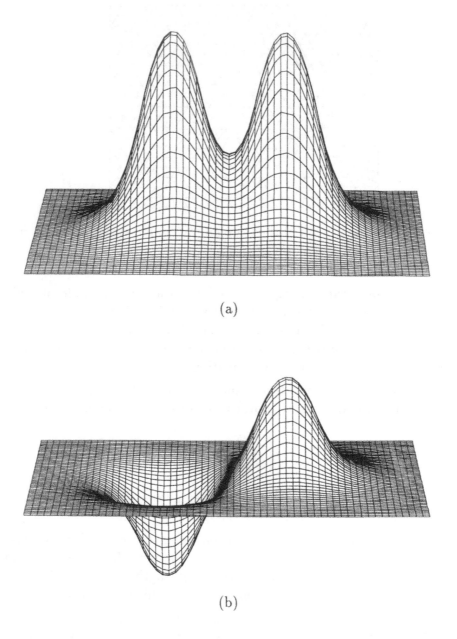

(a)

(b)

Figure 8.6: The modal electric field distributions of the two supermodes
in a linear directional coupler: (a) the even mode and (b) the odd mode.

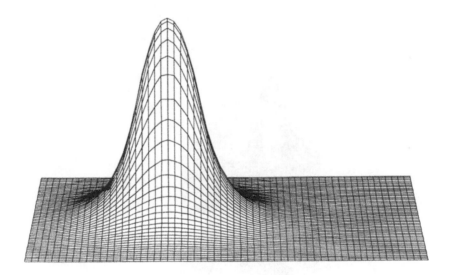

Figure 8.7: The modal electric field distribution of an isolated waveguide (Guide 1) in a linear directional coupler.

Figures 8.10 and 8.11, respectively. Now the coupling length is increased to $57\Delta\bar{z}$ approximately. When the length of the coupler is fixed, the optical power at the output can be switched between the two cores by properly choosing the input power. That is an all-optical switch. Moreover, when the nonlinear coupler is biased by a control beam that yields a balanced output, a probe beam input to the coupler could be amplified at the output. The result is an optical transistor.

To show the dependence of periodic optical switching on the total guided power, the powers guided in Guide 1, in the linear case and in the nonlinear case with $\widetilde{P} = 30.0$, are plotted in Figure 8.12 for comparison.

8.5.3 Application to spatial-temporal soliton systems

In §7.4 the propagation of spatial solitons was investigated. This section has shown how to apply propagation methods to coupled optical waveguides.

Another interesting research area is temporal soliton generation,

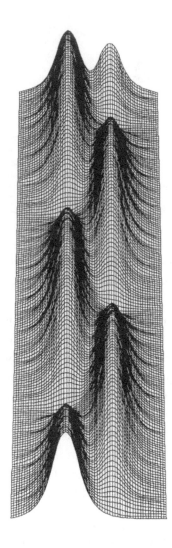

Figure 8.8: The propagation of a modal field distribution of an isolated waveguide in a linear directional coupler for 200 steps with $\Delta \bar{z} = 100.0$, where $F(\bar{x}, \bar{z})$ is plotted.

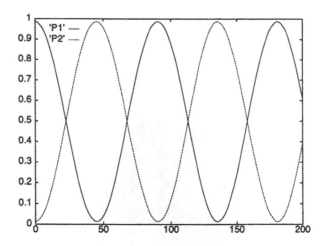

Figure 8.9: The periodic power switching in a linear directional coupler, where "P1" and "P2" are the guided powers in the individual waveguides in arbitrary units.

propagation and interaction in nonlinear optical waveguides, where use is made of both the nonlinearity and the dispersion of optical waveguides [350]-[361]. A temporal soliton, which has a hyperbolic-secant shape in time, is a stationary solution of the nonlinear Schrödinger equation. Physically there is a balancing between dispersion (pulse broadening) and self-focussing (pulse compression). Presently the transverse nonlinear effects are rarely considered when modelling temporal soliton systems. Such an approximation is acceptable for solitary pulses propagating in silica fibres whose nonlinear coefficient n_2 is only about 3.2×10^{-20} m^2/W [352]. Also the optical power from current semiconductor lasers is rather limited. These soliton propagation problems can be solved easily by using split-step Fourier methods [352]. With rapid advances in materials engineering and laser technology, optical materials having rather large nonlinear coefficients will turn the dream into reality sooner than later. By that time transverse nonlinear effects may become a dominant factor in the stability of bright and dark solitons propagating and interacting with one another in isolated as well as coupled waveguides [362]-[364]. Consequently, the FEM-FDM would be an obvious choice for studying the propagation of such spatial-temporal solitary pulses because of the flexibility of the FEM in coping with anisotropic and inhomogeneous media and because of the increased efficiency in computation achieved

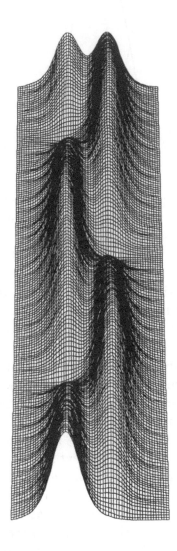

Figure 8.10: The propagation of a modal field distribution of an isolated waveguide in a nonlinear directional coupler for 200 steps with $\Delta \bar{z} = 100.0$ and $\widetilde{P} = 30.0$, where $F(\bar{x}, \bar{z})$ is plotted.

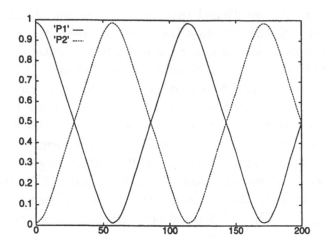

Figure 8.11: The periodic power switching in a nonlinear directional coupler, where "P1" and "P2" are the guided powers in the individual waveguides in arbitrary units with the total normalised power $\widetilde{P} = 30.0$.

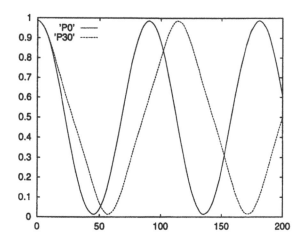

Figure 8.12: The dependence of periodic optical switching on the total guided power where "P0" and "P30" correspond, respectively, to the linear case (very low power) and to the nonlinear case with $\widetilde{P} = 30.0$. The vertical axis is in arbitrary units.

by using a graded mesh structure.

It should be mentioned that the applications of solitons are not restricted to high-capacity data transmission, where the particle-like nature of solitions makes them the natural bits for digital optical systems. Perhaps more significant applications of solitons arise in signal processing, such as soliton switching and logic. Moreover, when more powerful semiconductor lasers and new optical materials having large nonlinear coefficients come into being, an optical pulse can be compressed in both 3D-space and time domains, producing a spatial-temporal soliton in a saturable nonlinear optical waveguide. It is interesting to note that such a spatial-temporal soliton behaves like a light bullet [365, 366]. A nonlinear waveguide producing a train of such solitons may be termed an optical "machine-gun" [367], which could be triggered by a control soliton in a coupled waveguide structure.

It is rather unusual to end a book by telling the reader how to operate an optical "machine-gun", but we believe that the methodologies presented in this book will find their way into the modelling of spatial-temporal soliton systems and other nonlinear optical systems yet to be invented. We hope the reader has enjoyed reading the book and using the software.

Appendix A

ANOWS Contents

ANOWS stands for *Advanced Nonlinear Optical Waveguide Simulator*. This software package, though developed for nonlinear optical waveguides, can be applied equally to linear optical waveguides simply by setting to zero all nonlinear coefficients of permittivities of the waveguide to be analysed. It may also be applied to passive microwave waveguides, except those which support TEM modes.

The following is a list of the Fortran 77 programs contained in ANOWS, whose source code is given in the floppy diskette attached to this book.

AMG.FOR: Automatic Mesh Generation.

This program produces first- and second-order pre-graded triangular finite elements with smoothing and automatic bandwidth reduction.

OMG.FOR One-dimensional Mesh Generation.

This program produces second-order line elements.

MEF.FOR: Modal Analysis using the *E* Formulation.

Modal analysis of nonlinear optical waveguides by the finite element method using the full vectorial *E* formulation and second-order triangular elements.

The solution algorithm can be the conventional, the modified conventional or the non-conventional. Coordinates can be normalised with respect to either k_0 or β.

MHF.FOR Modal Analysis using the *H* Formulation.

Modal analysis of nonlinear optical waveguides by the finite element method using the vectorial *H* formulation and second-order triangular elements.

The solution algorithm can be the conventional, the modified conventional or the non-conventional. Coordinates can be normalised with respect to either k_0 or β.

MSA.FOR Modal Analysis using the Scalar *E*-Field **A**pproximation.

Modal analysis of nonlinear optical waveguides by the finite element method using the scalar *E*-field approximation and first-order triangular elements.

Dedicated to weakly-guiding structures or qualitative analysis and initial design.

The solution algorithm can be the conventional, the modified conventional or the non-conventional.

MTE.FOR Modal Analysis of **TE** Modes in Planar Structures.

Modal analysis of planar nonlinear optical waveguides by the finite element method using the scalar *E*-field and second-order line elements.

The solution algorithm can be the conventional, the modified conventional or the non-conventional.

MPP.FOR **M**esh **P**lot **P**rimer.

Generates data files for plotting a mesh by GNUPLOT, which is widely available.

FPP.FOR **F**ield **P**lot **P**rimer.

Generates data files for plotting the electric or magnetic field by GNUPLOT, which is widely available.

BMS.LIB **B**and **M**atrix **S**olver **LIB**rary.

A mathematical library containing a series of Fortran 77 subroutines for real banded matrix computation.

Note that not all Fortran programs used in this book are included in the above list. For example, the author has chosen not to release the nonlinear wave propagation program used in Chapters 7 and 8, because only a reflecting boundary was employed and an absorbing or transparent boundary might be preferable for general use.

Appendix B

ANOWS Documentation

B.1 AMG.FOR

This automatic mesh generation program produces first- and second-order pre-graded triangular finite elements with smoothing and automatic bandwidth reduction.

Input file

MESHIN.DAT: Contains the necessary information specifying the geometry of the structure to be investigated and is prepared by the user. The following is the one for Figure 4.9. The input is free format unless otherwise noted.

```
Data                                Comments
---------------------------- CUT HERE ---------------------
2,5,1                                 ; NR, NN, NA.
1,4,1,3.0,1,1                         ; RI, RT, MC, VNS, LNR.
2,3,2,8.0,2,3,4,5,1,1                 ; As above.
1,      0.0,        0.0,              ; NI, X, Y.
2,    -30.0,      -22.5,              ; As above.
3,     30.0,      -22.5,              ; As above.
4,     30.0,       22.5,              ; As above.
5,    -30.0,       22.5,              ; As above.
1,     10.0,                          ; AI, A.
---------------------------- CUT HERE ---------------------
```

```
where

 NR - Number of Regions;
 RI - Region Index (1 - NR);
 RT - Region Type (1 - 5);
 MC - Material Code (1,2,3,...);
VNS - Vertex Node Spacing within region;
LNR - List of Node and/or Radius Indices (as appropriate
for the particular type of region);
 NN - Number of Nodes used to define the whole structure
in this data file;
 NI - Node Index (1 - NN);
X,Y - Normalised Cartesian X and Y Coordinates of a Node;
 NA - Number of Radii;
 AI - Radius Index (1,2,3,...,NA);
  A - Normalised Radius.
```

and

- The first line begins with the number of regions into which the entire structure is to be divided. There are three reasons to decompose a geometrical structure into a number of regions: (a) to assign different material codes, (b) to use different vertex node spacings and (c) to make the structure compatible with the software (refer to Chapter 4). Also included are the total number of nodes and then the number of radii used for defining the various regions.

- The second and third lines are the list of regions providing information on the region type (one out of the five types outlined in Chapter 4), the material code (to be used in FEM programming), the vertex node spacing (for pre-graded mesh), and the list of relevant nodes and/or radii for the particular type of region. The number of such lines must equal the number of regions, the region index running consecutively from 1. Material codes are consecutive integers starting from 1, vertex node spacings are positive real numbers (acceptable by Fortran 77). For a region of

  Type 1: the 4 corner node indices are listed in counter-clockwise order;

  Type 2: the 4 corner node indices at the outer/inner boundary are listed in counter-clockwise/clockwise order with the nodes at the outer boundary being first;

Type 3: the 4 corner node indices are listed in counter-clockwise order followed by the centre node index and the radius index;

Type 4: the centre node index and then the radius index are listed;

Type 5: the centre node index and the radius index of the outer boundary are listed and then the radius index of the inner boundary.

- The fourth to the eighth lines are the list of nodes specifying the structure and their normalised Cartesian coordinates. The number of such lines must equal the total number of nodes, the node index running consecutively from 1. The interpretation of the normalisation is entirely up to the user as the program itself only accepts "numbers".

- The last lines comprise the list of radii, if any. The radius index runs consecutively from 1, and the radius value is normalised in the same way as the nodal coordinates.

Typical MESHIN.DAT files used in this book are given in the directory **MESHIN** of the floppy diskette attached. These files are named by their figure numbers with a prefix *MFIG*. For example, the MESHIN.DAT file for Figure 6.28 has the name MFIG6.28. To generate this mesh structure, the file MFIG6.28 is copied to MESHIN.DAT.

Interactive input from the terminal

- Select whether first- or second-order elements are to be generated.

- When selecting second-order elements, decide whether the associated first-order elements, obtained by decomposing each second-order element into four first-order ones, are also required.

These interactive inputs are self-explanatory.

Output files

ELMTR.DAT: Contains the number of elements, the semi-bandwidth of the mesh network, the order of the elements, and the list of elements. Each element has an element index, a material code and a list of the global nodes that it contains, in order of increasing local index number.

ELMT.DAT: Contains the above information on the first-order elements produced by the decomposition of second-order elements; optional.

NODRN.DAT: Contains the total number of global nodes, the list of nodes and their normalised coordinates.

The maximum number of nodes, elements, regions and bases (internal parameter for triangulation) are set to, respectively, 1000, 2000, 100 and 500. These parameters may be changed at the beginning of the program if the program itself prompts for an increase.

B.2   OMG.FOR

This one-dimensional mesh generation program produces second-order line elements for one-dimensional waveguide structures. The resulting data files are required by MTE.FOR for computing the TE modes of planar waveguides.

Interactive input from the terminal

- Specify the number of regions forming the structure. Normally, adjacent regions have different material code and/or node spacing.

- Specify the normalised coordinate of the starting (first) node.

- Specify the normalised coordinate of the end node and the node spacing for each region.

These interactive inputs are self-explanatory.

Output files

ELMT1.DAT: Contains the number of elements (first line) and the list of elements. Each element has an element index, a list of the global nodes that it contains (in order of the first node, the last node and the middle node), and a material code.

NOD1.DAT: Contains the total number of global nodes (first line), the list of nodes and their normalised coordinates.

The maximum number of elements is set to 500. This is believed to be more than necessary. Nevertheless, this parameter may be changed at the beginning of the program if desired.

B.3 MEF.FOR

This program has been developed for the modal analysis of nonlinear optical waveguides by the Galerkin finite element method using the full vectorial E formulation and second-order triangular elements.

The fundamental mode is extracted by the method of *Successive Over-relaxation* combined with the *Rayleigh Quotient*, whereas higher-order modes are computed by the method of *Bisection* and *Inverse Iteration with Shift*. Use is made of the banded structure for all matrix computations.

The solving of a system of linear equations by inverse iteration is achieved by means of LU decomposition followed by forward and back substitution. The LU decomposition is performed by using Gaussian elimination with partial pivoting.

Input files

ELMTR.DAT: Refer to Appendix B.1.

NODRN.DAT: Refer to Appendix B.1.

PRMTR.DAT: Data file of parameters and options. As an example the file used in § 6.4 is given below:

```
Parameters                 Comments
-------------- CUT HERE -------------------------
1.52         ; Used to form the penalty parameter.
Y            ; Normalisation option. Other option: N.
H            ; Material property. Other option: I.
1            ; Order of the mode desired.
-------------- CUT HERE -------------------------
```

where

- The first parameter is used to form the penalty parameter and is usually chosen as the refractive index of the substrate for the fundamental mode. The actual penalty parameter is its reciprocal for the *H* formulation and the reciprocal of its square for the *E* formulation. A higher-order mode requires a larger penalty parameter to suppress spurious modes. However, for the best accuracy of the computed eigenpair the penalty parameter chosen should be as small as possible.

- The normalisation option: "Y" for the structure to be normalised with respect to k_0, and "N" for the structure to be normalised with respect to β.

- Material property (referring to the linear term of the permittivity): "H" if the structure is piecewise homogeneous (region by region) and can be made compatible with the five types of regions described in Chapter 4; "I" if the structure must be treated as being entirely inhomogeneous.

- Order of the mode: "1" for the fundamental mode, "2" for the second lowest mode, and so on.

Typical PRMTR.DAT files used in this book are given in the directory **PRMTR** of the floppy diskette attached. These files are named by their section numbers with a prefix *PRM*. For example, the PRMTR.DAT file for § 6.4 has the name PRM6.4. To make use of it, the file PRM6.4 is copied to PRMTR.DAT.

SPNC.DAT: Contains a list of the saturation parameter α (the second column) and the nonlinear coefficients a (the third column) and b (the fourth column) for each *material code* (MC). Refer to Chapter 6 for the definitions of α, a and b. As an example, the file (free format) used in § 6.4 is given below.

```
MC              Data                        Comments
--------------------------- CUT HERE ------------------
1    0.00    0.00    0.00           ; Linear.
2    0.00    0.00    0.00           ; Linear.
3    1.00    1.00    1.00           ; Nonlinear.
4    0.00    0.00    0.00           ; Linear.
--------------------------- CUT HERE ------------------
```

Typical SPNC.DAT files used in this book are given in the directory **SPNC** of the floppy diskette attached. These files are named by their section numbers with a prefix *SPN*. For example, the SPNC.DAT file for Section 6.4 has the name SPN6.4. To make use of it, the file SPN6.4 is copied to SPNC.DAT.

DIEH.DAT: Contains the linear term of the relative permittivity dyadic for each homogeneous material. This free-format data file is required only when the material property "H" is selected in PRMTR.DAT.

The form of the relative permittivity dyadic is restricted to being hermitian and compatible with Equation 3.19. Again, "MC" stands for "Material Code". As an example, the file used in § 6.4 is given below.

MC	Row		Data (complex)	
		------- CUT HERE -------		
1	1	(2.3104,0.0000)	(0.0000,0.0000)	(0.0000,0.0000)
1	2	(0.0000,0.0000)	(2.3104,0.0000)	(0.0000,0.0000)
1	3	(0.0000,0.0000)	(0.0000,0.0000)	(2.3104,0.0000)
2	1	(2.4025,0.0000)	(0.0000,0.0000)	(0.0000,0.0000)
2	2	(0.0000,0.0000)	(2.4025,0.0000)	(0.0000,0.0000)
2	3	(0.0000,0.0000)	(0.0000,0.0000)	(2.4025,0.0000)
3	1	(2.3104,0.0000)	(0.0000,0.0000)	(0.0000,0.0000)
3	2	(0.0000,0.0000)	(2.3104,0.0000)	(0.0000,0.0000)
3	3	(0.0000,0.0000)	(0.0000,0.0000)	(2.3104,0.0000)
4	1	(1.0000,0.0000)	(0.0000,0.0000)	(0.0000,0.0000)
4	2	(0.0000,0.0000)	(1.0000,0.0000)	(0.0000,0.0000)
4	3	(0.0000,0.0000)	(0.0000,0.0000)	(1.0000,0.0000)
		------- CUT HERE -------		

Typical DIEH.DAT files used in this book are given in the directory **DIEH** of the floppy diskette attached. These files are named by their section numbers with a prefix *DIH*. For example, the DIEH.DAT file for Section 6.4 has the name DIH6.4. To make use of it, the file DIH6.4 is copied to DIEH.DAT.

DIEI.DAT: Contains the linear term of the relative permittivity dyadic for each element. This free-format data file is required only when the material property "I" is selected in PRMTR.DAT.

For each element, the values are listed on a node-by-node basis in terms of the *local* node index $(1, 2, \ldots, 6)$. This arrangement can also cope with abrupt jumps of the permittivity between adjacent elements, provided a different material code is used.

The user is expected to write a program to generate this data file as was done in § 6.3. Each line of this data file contains the following information:

element index, local node index, i, $\epsilon_r^l(i,1)$, $\epsilon_r^l(i,2)$, $\epsilon_r^l(i,3)$

where the row index i ($i = 1, 2, 3$) and $\epsilon_r^l(i,j)$ ($j = 1, 2, 3$) are the same as those in DIEH.DAT.

A practical file is usually lengthy, such as the one used in § 6.3, and will be omitted here.

RANDOM.DAT Contains random data as an initial field for the iteration. It supports a mesh structure of up to 500 nodes and is required only when the initial condition provided by the program itself is to be used. When a mesh has more than 500 nodes, it is necessary to enlarge the file by duplicating part of the data.

BIS.DAT: Contains the lower and upper bounds of the required eigenvalue in the matrix equation. This file is ignored when extracting the fundamental mode. The program prompts the user when the lower bound is set too high and/or the upper bound is set too low. However, for efficient computation the lower and upper bounds should be set as close to the desired eigenvalue as possible. The following is an example of BIS.DAT.

Lower Bound		Upper Bound
---------------------- CUT HERE	----------------------
0.1	20.0
---------------------- CUT HERE	----------------------

For an obvious reason the lower bound must be greater than zero.

Interactive input from the terminal and optional files

- Select the iteration algorithm: "P" if seeking the effective index β/k_0 for given normalised guided power $\widetilde{P}$; "B" if seeking $\widetilde{P}$ for given β/k_0. For the latter, the structure is restricted to being normalised with respect to k_0 as there is no computational advantage in normalising it with respect to β in this case.

- Specify either the normalised guided power $\widetilde{P}$ or the effective index β/k_0, depending on the above choice of iteration algorithm.

- Select an initial condition for the nonlinear iteration either from user input or from one provided by the program itself (RANDOM.DAT).

The initial condition from user input is assumed to be stored in files "EX.DAT", "EY.DAT" and "EZ.DAT", where these files are usually generated by a previous computation with different given power/effective-index.

The nonlinear iteration starts from a linear solution when the initial condition is provided by the program itself, that is, the nonlinear coefficient is set to zero for the first step of the nonlinear iteration. Otherwise, the permittivity is assumed to be nonlinear even for the first step. This arrangement increases the computational efficiency substantially.

The maximum number of nonlinear iterations is set to 50 in the program. If the nonlinear iteration procedure has not converged after 50 steps, it can be continued by restarting the program and choosing user input.

These interactive inputs are self-explanatory.

Output files

PWR.EIG: The normalised guided power $\widetilde{P}$ and the effective index β/k_0.

EN.DAT: The magnitude of the modal electric field, $\|E\|_2$.

EX.DAT: The x-component of the modal electric field, E_x.

EY.DAT: The y-component of the modal electric field, E_y.

EZ.DAT: The z-component of the modal electric field, $-jE_z$.

CONV.DAT: Convergence behaviour of the nonlinear iteration procedure: the iteration index i and the current eigenvalue λ_i for each step.

RSDL.DAT: The normalised guided power $\widetilde{P}$ and the relative residual r_r (refer to Chapter 5).

B.4   MHF.FOR

This program has been developed for the modal analysis of nonlinear optical waveguides by the Galerkin finite element method using the vectorial H formulation and second-order triangular elements. The function of this program is the same as that of MEF.FOR, including the input and output files, except

- that an eigenvector computed or used as an initial condition is a magnetic field stored in HN.DAT, HX.DAT, HY.DAT, and HZ.DAT rather than an electric field stored in EN.DAT, EX.DAT, EY.DAT, EZ.DAT in MEF.FOR; and

- that all off-diagonal terms of the permittivity tensor are restricted to be zero for MHF.FOR.

B.5   MSA.FOR

This program has been developed for the modal analysis of nonlinear optical waveguides by the Galerkin finite element method using the scalar E-field approximation and first-order triangular elements. It is dedicated to weakly-guiding structures or qualitative analysis and initial design. Here the structure is always normalised with respect to k_0, and the guided power is always normalised according to Equation 7.2.

Input files

ELMTR.DAT: Refer to Appendix B.1.

NODRN.DAT: Refer to Appendix B.1.

PRMTRS.DAT: Data file of parameters and options. The same PRMTRS.DAT file is used throughout the book (§ 7.2.1, § 7.2.2 and § 7.3) and is given below:

```
Parameters                              Comments
------------------ CUT HERE ----------------------
H                    ; Material property. Other option: I.
1                    ; Order of the mode desired.
------------------ CUT HERE ----------------------
```

where

- Material property (referring to the linear term of the permittivity): "H" if the structure is piecewise homogeneous (region by region) and can be made compatible with the five types of regions described in Chapter 4; "I" if the structure must be treated as being entirely inhomogeneous.

- Order of the mode: "1" for the fundamental mode, "2" for the second lowest mode, and so on.

This file is given in the directory **FEM.MSA** of the floppy diskette attached for convenience.

SPNC.DAT: Contains a list of the saturation parameter α (the first column) and the nonlinear coefficient a (the second column) for each *material code* (MC). Refer to Chapter 6 for the definitions of α and a. This file has the same name as the one used by MEF.FOR. The reason is that the file SPNC.DAT for MEF.FOR can actually be used by MSA.FOR since the program ignores the last column.

DIEH.DAT: Contains the linear term of the relative permittivity dyadic for each homogeneous material. This free-format data file is required only when the material property "H" is selected in PRMTRS.DAT.

This file has the same name and format as the one required by MEF.FOR for convenience. Only the first term of the relative permittivity dyadic for each material code is read by the program, the rest being ignored.

DIEI.DAT: Contains the linear term of the relative permittivity dyadic for each element. This free-format data file is required only when the material property "I" is selected in PRMTRS.DAT.

This file has the same name and format as the one required by MEF.FOR and MHF.FOR for convenience. Only the first term of the relative permittivity dyadic for each element is read by the program, the rest being ignored.

BIS.DAT: Contains the lower and upper bounds of the required eigenvalue in the matrix equation. This file is ignored when extracting the fundamental mode. The program prompts the user when the lower bound is set too high and/or the upper bound is set too low. This file has the same format as the one used by MEF.FOR.

Interactive input from the terminal and optional files

- Select the iteration algorithm: "P" if seeking the effective index β/k_0 for given normalised guided power $\widetilde{P}$; "B" if seeking $\widetilde{P}$ for given β/k_0.

- Specify either the normalised guided power $\widetilde{P}$ or the effective index β/k_0, depending on the above choice of iteration algorithm.

- Select an initial condition for the nonlinear iteration either from user input or from one provided by the program itself.

  The initial condition from user input is assumed to be stored in the file "ES.DAT", where this file is usually generated by a previous computation with different given power/effective-index.

  The nonlinear iteration starts from a linear solution when the initial condition is provided by the program itself, that is, the nonlinear coefficient is set to zero for the first step of the nonlinear iteration. Otherwise, the permittivity is assumed to be nonlinear even for the first step. This arrangement increases the computational efficiency substantially.

These interactive inputs are self-explanatory.

Output files

PWR.EIG: The normalised guided power $\tilde{P}$ and the effective index β/k_0.

ES.DAT: The modal electric field (scalar), E.

CONV.DAT: Convergence behaviour of the nonlinear iteration procedure: the iteration index i and the current eigenvalue λ_i for each step.

RSDL.DAT: The normalised guided power $\tilde{P}$ and the relative residual r_r (refer to Chapter 5).

B.6 MTE.FOR

This program has been developed for the modal analysis of planar non-linear optical waveguides by the Galerkin finite element method using the scalar E-field and second-order line elements. It is dedicated to computing linear and nonlinear TE modes in planar structures. Here the structure is always normalised with respect to k_0. The guided power per unit length is normalised according to § 6.5.

Input files

ELMT1.DAT: Refer to Appendix B.2.

NOD1.DAT: Refer to Appendix B.2.

PRMTR1.DAT: Data file of parameters and options. As an example, the file used in § 6.5 is given below:

```
Parameters                              Comments
------------------- CUT HERE  --------------------------
H               ; Material property. Other option: I.
1               ; Order of the mode desired.
------------------- CUT HERE  --------------------------
```

where

- Material property (referring to the linear term of the permittivity): "H" if the structure is piecewise homogeneous (layer by layer); "I" if the structure must be treated as being entirely inhomogeneous (reserved for future use).

  Note that the latter option has not yet been incorporated into the program, because such a situation is rare. At present, users can treat an inhomogeneous problem as a piecewise homogeneous one on an element by element basis. The error resulting from the approximation will be negligible when element sizes are small.

- Order of the mode: "1" for the fundamental mode, "2" for the second mode, and so on.

This file is virtually the same as the PRMTRS.DAT in Appendix B.5 except that the interpretations are slightly different.

DIE1.DAT: Contains a list of the linear term of the relative permittivity (second column), the normalised nonlinear coefficient (third column), and the saturation parameter (fourth column) for each *material code* (MC). This free-format data file is required when the material property "H" is selected in PRMTR1.DAT. As an example, the file used in § 6.5 is given below:

```
MC     LP      NC      SP              Comments
------------------------ CUT HERE ------------------------
1     2.4649   0.0     0.0     ;Linear thin-film.
2     2.4025   1.0     1.0     ;Nonlinear bounding layers.
------------------------ CUT HERE ------------------------
```

where "MC","LP","NC", and "SP" represent material code, linear permittivity, nonlinear coefficient, and saturation parameter.

BIS.DAT: Contains the lower and upper bounds of the required eigenvalue in the matrix equation. This file is ignored when extracting the fundamental mode. The program prompts the user when the lower bound is set too high and/or the upper bound is set too low. This file is the same as the one used by MEF.FOR.

Interactive input from the terminal and optional files

- Select the iteration algorithm: "P" if seeking the effective index β/k_0 for given normalised guided power per unit length $\tilde{p}$; "B" if seeking $\tilde{p}$ for given β/k_0.

- Specify either the normalised guided power per unit length or the effective index β/k_0, depending on the above choice of iteration algorithm.

- Select an initial condition for the nonlinear iteration either from user input or from one provided by the program itself.

  The initial condition from user input is assumed to be stored in the files "E1.DAT" and "PWR.EIG", where these files are usually generated by a previous computation with different given power/ effective-index.

  The nonlinear iteration starts from a linear solution when the initial condition is provided by the program itself, that is, the nonlinear coefficient is set to zero for the first step of the nonlinear iteration. Otherwise, the permittivity is assumed to be nonlinear even for the first step. This arrangement increases the computational efficiency substantially.

These interactive inputs are self-explanatory.

Output files

PWR.EIG: The normalised guided power per unit length $\tilde{p}$ and the effective index β/k_0.

E1.DAT: The modal electric field (scalar and one-dimensional), E.

CONV.DAT: Convergence behaviour of the nonlinear iteration procedure: the iteration index i and the current eigenvalue λ_i for each step.

RSDL.DAT: The normalised guided power per unit length $\tilde{p}$ and the relative residual r_r (refer to Chapter 5).

B.7   MPP.FOR

Generate data files for plotting a mesh by GNUPLOT, which is widely available.

The maximum number of nodes is set to 1000, and the maximum number of elements is set to 2000 in the program. This is believed to be sufficient for most applications. Nevertheless, the user can change these parameters at the beginning of the program if desired.

Input files

ELMTR.DAT: Refer to Appendix B.1.

NODRN.DAT: Refer to Appendix B.1.

Output files

MSH.PLT: Data file for plotting a triangular mesh by GNUPLOT, where this file must be plotted with the *lines* or *linespoints* option for the *data style*.

NOD.PLT: Data file for plotting the nodes of a mesh by GNUPLOT, where this file must be plotted with the *points* option for the *data style*.

B.8   FPP.FOR

Generate data files for plotting the electric or magnetic field by GNU-PLOT, which is widely available.

In the program the maximum number of nodes and the maximum number of elements are set to 1000. The maximum number of data points that can be generated in the x- and y-directions is set to 200 for each direction. This is believed to be sufficient for most applications. Nevertheless, the user can change these parameters at the beginning of the program if desired.

Input files

ELMTR.DAT: Refer to Appendix B.1.

NODRN.DAT: Refer to Appendix B.1.

PLOT.IN: Data file of a scalar field or one component of the electric or magnetic vector field that is to be plotted.

This file can be a copy of ES.DAT, EN.DAT, EX.DAT, EY.DAT, EZ.DAT, HN.DAT, HX.DAT, HY.DAT, or HZ.DAT.

Interactive input from the terminal

Select whether or not to change the sign of the field values to be plotted.

Output files

PLOT.OUT: Generated data file to be plotted by GNUPLOT.

B.9 BMS.LIB

BMS.LIB is a mathematical library containing a series of Fortran 77 subroutines for banded matrix computation. These subroutines are self-explanatory. They are used by and have been appended to MEF.FOR, MHF.FOR, MSA.FOR and MTE.FOR. Thus, users need no further action in using MEF.FOR, MHF.FOR, MSA.FOR and MTE.FOR. The following is a brief description of these subroutines.

RELAXATION

RELAXATION is based on the *Successive Over-Relaxation* (SOR) together with *Rayleigh quotient* to find the eigenpair corresponding to the lowest eigenvalue of the matrix equation: $Ax = \lambda Bx$, where A and B are real symmetric matrices stored in banded form. Note that RELAXATION can be used to find the eigenpair corresponding to the largest eigenvalue by rewriting the matrix equation in the form $Bx = \eta Ax$ and taking $\eta = 1/\lambda$ as the new eigenvalue.

Generally speaking, RELAXATION has global convergence. Because it is iterative in nature, the initial eigenvector for the iteration may have a great bearing on the convergence. For example, when the initial eigenvector is much closer to the eigenvector corresponding to the second lowest eigenvalue and this second lowest eigenvalue is almost degenerate to the lowest eigenvalue, RELAXATION may yield the second lowest eigenpair. This happens only to MEF.FOR and MHF.FOR when the lowest x-polarised mode and the lowest y-polarised mode are almost degenerate. Also, when the initial eigenpair for the iteration is almost exactly a higher-order mode, RELAXATION may simply reproduce this mode. This is because the function of the *Rayleigh quotient* technique is to find the stationary value of the functional:

$$F(x) = \frac{x^\dagger A x}{x^\dagger B x} \qquad (B.1)$$

and this functional has a stationary value for such an eigenpair. The above is just a word of warning. The random data used as the initial condition for MEF.FOR and MHF.FOR should yield the results desired, or at least one of the lowest x- and y-polarised modes in the case when they are almost degenerate. It is always safe to check whether the computed solution is the fundamental mode by simply interchanging the x and y components of the computed solution and using the new field as an initial condition to do an extra computation. If the new eigenvalue is lower, then the new field is the fundamental mode. The fundamental mode is used as the initial condition for all subsequent computations at higher powers.

BISECTION

BISECTION is developed for finding the nth lowest eigenvalue of the matrix equation: $Ax = \lambda Bx$ using the *Sturm Sequence Property* and repeated bisection, where A and B are real symmetric banded matrices with B being positive definite. It can also be used to find the nth highest eigenvalue using the same technique as discussed for RELAXATION.

In the ANOWS package BISECTION is used only for extracting the higher-order modes because of its intensive computation.

ITERATION

ITERATION is built on *inverse iteration* with or without shift. It is developed for finding the eigenpair whose eigenvalue is closest to the estimated eigenvalue given by BISECTION and is required only for computing higher-order modes.

ITERATION has an argument ISHF. The *inverse iteration* is performed without shift when ISHF is set to zero and with shift when ISHF is set to 1.

The purpose of *Inverse iteration* without shift is to find the eigenvector corresponding to the smallest eigenvalue and has global convergence; and *inverse iteration* with shift is used to find the eigenvector corresponding to the eigenvalue closest to the estimated eigenvalue given and has local convergence. ITERATION also uses the *Rayleigh quotient* technique for eigenpair refinement.

It is noticed that the convergence of the inverse iteration without shift is much slower than that of the successive over-relaxation. Accordingly, the latter is adopted for computing the fundamental mode.

The *inverse iteration* is performed by LU decomposition followed by forward and back substitution.

For the inverse iteration with shift the matrix is nearly singular when the estimated eigenvalue approaches the true eigenvalue. Consequently, the convergence of the iteration may be slow when the initial eigenvector is far from the true eigenvector such as when using random data. This happens only to the solution of the first eigenpair on a power dispersion curve because a good initial eigenvector is always available for subsequent computations.

LUFACT

LUFACT performs the LU decomposition of a real banded matrix stored in band storage mode by Gaussian elimination with partial pivoting. It is required only for finding higher-order modes by BISECTION and ITERATION.

AXBSOL

AXBSOL is developed to solve the real banded matrix system equation $Ax = b$ using the LU factors computed by LUFACT. It is required only for finding higher-order modes by ITERATION.

DETERM

DETERM computes the determinant of a real banded matrix using the LU factors computed by LUFACT. It is required only for finding higher-order modes by BISECTION.

RESIDUAL

RESIDUAL evaluates the relative residual of the generalised eigenvalue equation $Ax = \lambda Bx$ for a computed eigenpair. It uses pre-computed results from RELAXATION or ITERATION, depending on the order of the mode sought. Refer to Chapter 5 for the algorithm and definition.

Appendix C

ANOWS Usage and Examples

In the attached floppy diskette those directories holding one or more Fortran 77 source files are self-contained for testing. That is, the user can simply generate executable code from that source code and run the executable code without involving another directory or preparing any other data file. The generated mesh or computed field can also be plotted from within the same directory. It is highly recommended that the user starts with the triangular mesh generation and scalar field computation and then moves to the vector field computation, which is much more intensive in computation.

C.1 Mesh Generation and Plotting

The one-dimensional mesh generation and plotting can be understood easily and are omitted here. In the following the term "mesh" is assumed to be two-dimensional unless noted otherwise.

Mesh generation

1. Prepare an input data file: MESHIN.DAT;

2. Generate the executable code from the source code: AMG.FOR, using the system FORTRAN 77 compiler;

3. Run the executable code and make selections from the terminal (self-explanatory).

The program automatically generates the output data files ELMTR.DAT, NODRN.DAT and optionally ELMT.DAT.

An example of MESHIN.DAT and the associated ELMTR.DAT, ELMT.DAT and NODRN.DAT are given in the directory **MESH** of the attached floppy diskette.

Mesh plotting

1. Obtain the input data files ELMTR.DAT and NODRN.DAT from the automatic mesh generation program AMG.FOR;

2. Generate the executable code from the source code: MPP.FOR, using the system FORTRAN 77 compiler;

3. Run the executable code to produce the output data files MSH.PLT and NOD.PLT.

The data files MSH.PLT and NOD.PLT can be plotted by GNUPLOT in the following way. Within the GNUPLOT shell

- set terminal *user terminal type*

- set nokey

- set nozeroaxis

- set noborder

- set noxtics

- set noytics

- plot 'MSH.PLT' with lines, 'NOD.PLT' with points

- set terminal postscript

- set output 'MESH.PS'

- replot

- quit

The output postscript file MESH.PS contains the mesh structure and can be printed by a laser printer. The "plot" command in the 7th statement enables the user to see the mesh structure on the terminal. The plotting of the file NOD.PLT may be omitted if only the mesh is required. An example of MSH.PLT and NOD.PLT is given in the directory **MESH** of the attached floppy diskette.

The initialisation of GNUPLOT may be different from computer to computer. Other options may be set to suit individual needs. For more information the GNUPLOT manual or online help should be consulted.

C.2   Vector Modal Analysis

MEF.FOR

1. Obtain the input data files ELMTR.DAT and NODRN.DAT from the automatic mesh generation program AMG.FOR;

2. Prepare data files: PRMTR.DAT, SPNC.DAT, DIEH.DAT/ DIEI.DAT, and optionally BIS.DAT if a higher-order mode is sought;

3. Make the file RANDOM.DAT available or the files EX.DAT, EY.DAT and EZ.DAT available if selecting the initial condition for the nonlinear iteration from user input;

4. Generate the executable code from the source code: MEF.FOR, using the system FORTRAN 77 compiler;

5. Run the executable code and select the nonlinear iteration algorithm: "P" (seeking the effective index for given normalised guided power) or "B" (seeking the normalised guided power for given effective index);

6. Specify either the normalised guided power $\widetilde{P}$ or the effective index β/k_0, depending on the above choice of iteration algorithm;

7. Select the initial condition for the nonlinear iteration either from user input or from the one provided by the program itself.

The computed results will be written in files: PWR.EIG, EN.DAT, EX.DAT, EY.DAT, EZ.DAT, RSDL.DAT and CONV.DAT.

Examples of input files: ELMTR.DAT, NODRN.DAT, PRMTR.DAT, SPNC.DAT, DIEH.DAT, EN.DAT, EX.DAT, EY.DAT, EZ.DAT, BIS.DAT and RANDOM.DAT are given in the directory **FEM.VEC** of the attached floppy diskette. Note that the file RANDOM.DAT supports up to 500 nodes.

MHF.FOR

The program MHF.FOR can be treated in the same way as MEF.FOR provided that the files EN.DAT, EX.DAT, EY.DAT and EZ.DAT are replaced by HN.DAT, HX.DAT, HY.DAT and HZ.DAT. Examples of input files: HN.DAT, HX.DAT, HY.DAT and HZ.DAT are also given in the directory **FEM.VEC** of the attached floppy diskette.

C.3   Scalar E-Field Approximation

MSA.FOR

1. Obtain the input data files ELMTR.DAT and NODRN.DAT from the automatic mesh generation program AMG.FOR;

2. Prepare data files: PRMTRS.DAT, SPNC.DAT, DIEH.DAT/ DIEI.DAT, and optionally BIS.DAT if a higher-order mode is sought;

3. Make the file ES.DAT available if selecting the initial condition for the nonlinear iteration from user input;

4. Generate the executable code from the source code: MSA.FOR, using the system FORTRAN 77 compiler;

5. Run the executable code and select the nonlinear iteration algorithm: "P" (seeking the effective index for given normalised guided power) or "B" (seeking the normalised guided power for given effective index);

6. Specify either the normalised guided power $\tilde{P}$ or the effective index β/k_0, depending on the above choice of iteration algorithm;

7. Select the initial condition for the nonlinear iteration either from user input or from the one provided by the program itself.

The computed results will be written in files: PWR.EIG, ES.DAT and CONV.DAT.

Examples of input files: ELMTR.DAT, NODRN.DAT, ES.DAT, PRMTRS.DAT, DIEH.DAT, SPNC.DAT and BIS.DAT are given in the directory **FEM.MSA** of the attached floppy diskette.

C.4 Analysis of TE Modes in Planar Structures

MTE.FOR

1. Generate the input data files ELMT1.DAT and NOD1.DAT using the one-dimensional mesh generation program OMG.FOR;

2. Prepare data files: PRMTR1.DAT, DIE1.DAT, and optionally BIS.DAT if a higher-order mode is sought;

3. Make the files E1.DAT and PWR.EIG available if selecting the initial condition for the nonlinear iteration from user input;

4. Generate the executable code from the source code: MTE.FOR, using the system FORTRAN 77 compiler;

5. Run the executable code and select the nonlinear iteration algorithm: "P" (seeking the effective index for given normalised guided power per unit length) or "B" (seeking the normalised guided power for given effective index);

6. Specify either the normalised guided power per unit length $\tilde{p}$ or the effective index β/k_0, depending on the above choice of iteration algorithm;

7. Select the initial condition for the nonlinear iteration either from user input or from the one provided by the program itself.

The computed results will be written in files: PWR.EIG, E1.DAT and CONV.DAT.

Examples of input files: ELMT1.DAT, NOD1.DAT, PRMTR1.DAT, BIS.DAT, PWR.EIG, DIE1.DAT and E1.DAT are given in the directory **FEM.MTE** of the attached floppy diskette.

Note that the file E1.DAT can be plotted by GNUPLOT. The plotting procedure is straightforward and is omitted for brevity.

C.5   Field Plot Primer

FPP.FOR

1. Copy the file to be plotted to the file PLOT.IN;

2. Make the associated data files ELMTR.DAT and NODRN.DAT available;

3. Generate the executable code from the source code: FPP.FOR, using the system FORTRAN 77 compiler;

4. Run the executable code to produce the output data file PLOT.OUT.

The file to be plotted can be one of the following files: ES.DAT, EN.DAT, EX.DAT, EY.DAT, EZ.DAT, HN.DAT, HX.DAT, HY.DAT, and HZ.DAT. Examples of these files are given in the directory **FEM.MSA** and **FEM.VEC** of the attached diskette. The associated files ELMTR.DAT and NODRN.DAT are also given in the relevant directory.

PLOT.OUT can be plotted by GNUPLOT in the following way. Within the GNUPLOT shell

- set terminal *user terminal type*

- set data style lines

- set noparametric

- set hidden3d

- set noxtics

- set noytics

- set noborder

- set nokey

- splot 'PLOT.OUT' using 3

- set terminal postscript

- set output 'PLOT.PS'

- replot

- quit

The output postscript file PLOT.PS contains the field and can be printed by a laser printer. The "splot" command in the 9th statement enables the user to see the three-dimensional field plot on the terminal. An example of PLOT.OUT is given in the directory **FEM.VEC**.

The initialisation of GNUPLOT may be different from computer to computer. Other options may be set to suit individual needs. Note that GNUPLOT also supports contour plot. For more information the GNUPLOT manual or online help should be consulted.

Appendix D

Common Errors in Using ANOWS

- **Compiling error**

  Certain matrices are too big or poorly declared. This may happen when using a PC compiler because the matrix sizes set in the program are larger than the maximum sizes permitted by the compiler. To solve the problem the program should be moved to a main frame computer or the matrix sizes reduced (at the beginning of the program).

  Increase certain parameters. The program may prompt the user to increase certain parameters because the default matrix sizes set in the program are not big enough for the structure being analysed. Such parameters can be changed easily at the beginning of the program.

- **Executing error**

  Cannot execute the program. This may happen when using a PC. The reason and remedy are the same as those for the *compiling error*.

- **Input file error**

  Certain files do not exist or *cannot open certain files.* The required input files are not available to the computer.

  End of file reached. Poor input file with insufficient data.

- **Output file error**

Certain files already exist or *cannot open certain files*. The *STATUS* of output files are set to *NEW* in the program. If the operating system allows the user neither to overwrite an existing file nor to have more than one copy of the same named file, then such an error will arise.

The existing output files may be deleted or renamed. The default system setup could be changed.

- **Meaningless results**

Poor input files or non-physical parameters were used.

Spurious modes have appeared because of an inappropriate penalty parameter.

- **The eigenpair solved is not the desired mode**

This may happen when the structure has degenerate or nearly degenerate modes. There is no known strategy that can completely solve the problem though certain techniques are well worth trying, such as slightly perturbing the initial condition for the iteration or choosing a favourable initial condition for the desired mode.

This happens only when the structure possesses a certain symmetry property or when the x- and y-polarised modes have almost identical eigenvalues.

Here "nearly degenerate" should be interpreted as the difference in the eigenvalues being much less than a fraction of one percent.

- **The iteration procedure does not converge but oscillates**

This happens to higher-order nonlinear modes in certain structures such as the one presented in §6.5. For example, if one tries to seek the second-order nonlinear scalar mode, such a waveguide will behave like a nonlinear coupler. Unlike a linear coupler, which has stationary solutions called supermodes or array modes, a coupler under strong nonlinear action does not possess such stationary solutions, and therefore modal analyses do not apply to such a structure. Nonlinear couplers should be analysed by nonlinear coupled-mode theories or wave/beam propagation methods, which are more accurate.

Bibliography

[1] N.L. Boling, A.J. Glass and A. Owyoung, "Empirical relationships for predicting nonlinear refractive index changes in optical solids," IEEE J. Quantum Electron. **QE-14,** 601-608 (1978).

[2] R. Adair, L.L. Chase and S.A. Payne, "Nonlinear refractive-index measurements of glass using three-wave frequency mixing," J. Opt. Soc. Am. B **4,** 875-881 (1987).

[3] G.I. Stegeman and R.H. Stolen, "Waveguides and fibers for nonlinear optics," J. Opt. Soc. Am. B **6,** 652-662 (1989).

[4] G.I. Stegeman, "Material figures of merit and implications to all-optical waveguide switching," Proceedings of the SPIE **1852,** 75-89 (1993).

[5] R. Burzynski, B.P. Singh, P.N. Prasad, R. Zanoni and G.I. Stegeman, "Nonlinear optical processes in a polymer waveguide: Grating coupler measurements of electronic and thermal nonlinearities," Appl. Phys. Lett. **53,** 2011-2013 (1988).

[6] L. Yang *et al.*, "Third-order optical nonlinearity in polycondensed thiophene-based polymers and polysilane polymers," J. Opt. Soc. Am. B **6,** 753-756 (1989).

[7] B.M.A. Rahman, F.A. Fernandez and J.B. Davies, "Review of finite element methods for microwave and optical waveguides," Proceedings of the IEEE **79,** 1442-1448 (1991).

[8] A.W. Snyder and J.D. Love, **Optical waveguide theory,** London: Chapman and Hall, 1983.

[9] A. Konrad, "Higher-order triangular finite elements for electromagnetic waves in anisotropic media," IEEE Trans. Microwave Theory Tech. **MTT-25,** 353-360 (1977).

[10] B.M.A. Rahman and J.B. Davies, "Finite-element analysis of optical and microwave waveguide problems," IEEE Trans. Microwave Theory Tech. **MTT-32,** 20-28 (1984).

[11] M. Koshiba, K. Hayata and M. Suzuki, "Finite-element solution of anisotropic waveguides with arbitrary tensor permittivity," J. Lightwave Technol. **LT-4,** 121-126 (1986).

[12] J.P. Webb, "Finite element analysis of dispersion in waveguides with sharp metal edges," IEEE Trans. Microwave Theory Tech. **MTT-36,** 1819-1824 (1988).

[13] W.J. English and F.J. Young, "An E vector variational formulation of the Maxwell equations for cylindrical waveguide problems," IEEE Trans. Microwave Theory Tech. **MTT-19,** 40-46 (1971).

[14] M. Koshiba, K. Hayata and M. Suzuki, "Finite-element formulation in terms of the electric field vector for electromagnetic waveguide problems," IEEE Trans. Microwave Theory Tech. **MTT-33,** 900-905 (1985).

[15] O. Picon, "Three-dimensional finite-element formulation for deterministic waveguide problems," Microwave Opt. Technol. Lett. **1,** 170-172 (1988).

[16] J.A.M. Svedin "A numerically efficient finite-element formulation for the general waveguide problem without spurious modes," IEEE Trans. Microwave Theory Tech. **MTT-37,** 1708-1715 (1989).

[17] C. Yeh, K. Ha, S.B. Dong and W.P. Brown, "Single-mode optical waveguides," Appl. Opt. **18,** 1490-1504 (1979).

[18] K. Oyamada and T. Okoshi, "Two-dimensional finite-element method calculation of propagation characteristics of axially non-symmetrical optical fibers," Radio Science **17,** 109-116 (1982).

[19] D. Gelder, "Numerical determination of microstrip properties using the transverse field components," Proc. IEE **117,** 699-703 (1970).

[20] C.G. Williams and G.K. Cambrell, "Numerical solution of surface waveguide modes using transverse field components," IEEE Trans. Microwave Theory Tech. **MTT-22,** 329-330 (1974).

[21] K. Hayata, M. Eguchi and M. Koshiba, "Finite element formulation for guided-wave problems using transverse electric field component," IEEE Trans. Microwave Theory Tech. **MTT-37,** 256-258 (1989).

[22] W.C. Chew and M.A. Nasir, "A variational analysis of anisotropic, inhomogeneous dielectric waveguides," IEEE Trans. Microwave Theory Tech. **MTT-37,** 661-668 (1989).

[23] K. Hayata, M. Nagai and M. Koshiba, "Finite-element theory of nonlinear TM-polarised guided waves," Electron. Lett. **23,** 1305-1307 (1987).

[24] K. Hayata, M. Nagai and M. Koshiba, "Finite-element formalism for nonlinear slab-guided waves," IEEE Trans. Microwave Theory Tech. **36,** 1207-1215 (1988).

[25] K. Hayata and M. Koshiba, "Full vectorial analysis of nonlinear-optical waveguides," J. Opt. Soc. Am. B **5,** 2494-2501 (1988).

[26] P. Bettess, "Finite element modelling of exterior electromagnetic problems," IEEE Trans. Magn. **MAG-24,** 238-243 (1988).

[27] C.G. Williams and G.K. Cambrell, "Efficient numerical solution of unbounded field problems," Electron. Lett. **8,** 247-248 (1972).

[28] B.H. McDonald and A. Wexler, "Finite-element solution of unbounded field problems," IEEE Trans. Microwave Theory Tech. **MTT-20,** 841-847 (1972).

[29] S. Washisu, I. Fukai and M. Suzuki, "Extension of finite-element method to unbounded field problems," Electron. Lett. **15,** 772-774 (1979).

[30] C.C. Su, "A combined method for dielectric waveguides using the finite-element technique and the surface integral equations method," IEEE Trans. Microwave Theory Tech. **MTT-34,** 1140-1146 (1986).

[31] J.F. Lee and Z.J. Cendes, "Transfinite elements: A highly efficient procedure for modelling open field problems," J. Appl. Phys. **61,** 3913-3915 (1987).

[32] P. Bettess, "More on infinite elements," Int. J. Numer. Methods Eng. **15**, 1613-1626 (1980).

[33] O.C. Zienkiewicz, C. Emson and P. Bettess, "A novel boundary infinite element," Int. J. Numer. Methods Eng. **19**, 393-404 (1983).

[34] F. Medina and R.L. Taylor, "Finite element techniques for problems of unbounded domains," Int. J. Numer. Methods Eng. **19**, 1209-1226 (1983).

[35] K. Hayata, M. Eguchi and M. Koshiba, "Self-consistent finite/infinite element scheme for unbounded guided wave problems," IEEE Trans. Microwave Theory Tech. **MTT-36**, 614-616 (1988).

[36] M.J. McDougall and J.P. Webb, "Infinite elements for the analysis of open dielectric waveguides," IEEE Trans. Microwave Theory Tech. **MTT-37**, 1724-1731 (1989).

[37] Y.R. Shen, **The principles of nonlinear optics**, New York: John Wiley & Sons, 1984.

[38] A.D. Boardman, A.A. Maradudin, G.I. Stegeman, T. Twardowski and E.M. Wright, "Exact theory of nonlinear p-polarized optical waves," Phys. Rev. A **35**, 1159-1164 (1987).

[39] L.D. Landau and E.M. Lifshitz, **Electrodynamics of continuous media,** 2nd edn., Oxford: Pergamon Press, 1984.

[40] U. Langbein, F. Lederer, T. Peschel and H.-E. Ponath, "Nonlinear guided waves in saturable nonlinear media," Opt. Lett. **10**, 571-573 (1985).

[41] G.I. Stegeman *et al.,* "Nonlinear slab-guided waves in non-Kerr-like media," IEEE J. Quantum Electron. **QE-22**, 977-983 (1986).

[42] P.M. Lambkin and K.A. Shore, "Asymmetric semiconductor waveguides with defocussing nonlinearity," IEEE J. Quantum Electron. **QE-24**, 2046-2051 (1988).

[43] S.J. Al-Bader and H.A. Jamid, "Nonlinear waves in saturable self-focusing thin films bounded by linear media," IEEE J. Quantum Electron. **QE-24**, 2052-2058 (1988).

[44] R.F. Harrington, "The method of moments in electromagnetics," J. Electromagnetic Waves and Applications **1**, No. 3, 181-200 (1987).

[45] G.K. Cambrell, " A short course on **the finite element analysis of electromagnetic devices,** " Course Notes, Monash University, Melbourne, Australia, 1988.

[46] R.F. Harrington, **Field computation by moment methods,** New York: Macmillan, 1968.

[47] M. Koshiba, K. Hayata and M. Suzuki, "Vector E-field finite-element analysis of dielectric optical waveguides," Appl. Opt. **25,** 10-11 (1986).

[48] J.P. Webb, "Efficient generation of divergence-free fields for the finite element analysis of 3D cavity resonances," IEEE Trans. Magn. **MAG-24,** 162-165 (1988).

[49] A.R. Pinchuk, C.W. Crowley, P.P. Silvester, and R.L. Ferrari, "Spectrally correct finite element operators for electromagnetic vector fields," J. Appl. Phys. **63,** 3025-3027 (1988).

[50] S.H. Wong and Z.J. Cendes, "Combined finite element-modal solution of three-dimensional eddy current problems," IEEE Trans. Magn. **MAG-24,** 2685-2687 (1988).

[51] J.F. Lee, D.K. Sun and Z.J. Cendes, "Full-wave analysis of dielectric waveguides using tangential vector finite elements," IEEE Trans. Microwave Theory Tech. **39,** 1262-1271 (1991).

[52] A. Konrad, "A method for rendering 3D finite element vector field solutions non-divergent," IEEE Trans. Magn. **MAG-25,** 2822-2824 (1989).

[53] S.H. Wong and Z.J. Cendes, "Numerically stable finite element methods for the Galerkin solution of eddy current problems," IEEE Trans. Magn. **MAG-25,** 3019-3021 (1989).

[54] M. Koshiba, K. Hayata and M. Suzuki, "Study of spurious solutions of finite-element method in the three-component magnetic field formulation for dielectric waveguide problems," Electron. Commun. Japan, Part 1, **68,** 114-119 (1985).

[55] A.R. Pinchuk, C.W. Crowley and P.P. Silvester, "Spurious solutions to vector diffusion and wave field problems," IEEE Trans. Magn. **MAG-24,** 158-161 (1988).

[56] J.P. Aubin, **Approximation of elliptic boundary-value problems,** New York: Wiley-Interscience (John Wiley & Sons), 1972.

[57] J.T. Oden and J.N. Reddy, **An introduction to the mathematical theory of finite elements,** New York: Wiley-Interscience (John Wiley & Sons), 1976.

[58] J.T. Oden and J.N. Reddy, **Variational methods in theoretical mechanics,** Berlin: Springer-Verlag, 1976.

[59] B. Friedman, **Principles and techniques of applied mathematics,** New York: John Wiley & Sons, 1956, 148-150.

[60] G.F. Carey and J.T. Oden, **Finite elements: a second course,** Vol. II, Englewood Cliffs, N.J.: Prentice-Hall, 1983.

[61] J.T. Oden and G.F. Carey, **Finite elements: mathematical aspects,** Vol. IV, Englewood Cliffs, N.J.: Prentice-Hall, 1983.

[62] G. Bachman and L. Narici, **Functional analysis,** New York: Academic Press, 1966.

[63] A.W. Naylor and G.R. Sell, **Linear operator theory in engineering and science,** New York: Holt, Rinehart and Winston, 1971.

[64] C. Lanczos, "Boundary value problems and orthogonal expansions," SIAM J. Appl. Math. **14,** 831-863 (1966).

[65] G.K. Cambrell, **Linear operators and variational principles in electromagnetic theory,** Ph.D. Thesis, Monash University, Melbourne, Australia, 1972.

[66] G.K. Cambrell and G.C. Williams, "Non-self-adjoint operators in waveguide problems," Digest of Int. Electron. Convention, Australia, 319-321 (1975).

[67] J.B. Davies, "Finite element analysis of waveguides and cavities-a review," IEEE Trans. Magnet. **29,** 1578-1583 (1993).

[68] M.L. Barton and Z.L. Cendes, "New vector finite elements for three-dimensional magnetic field computation," J. Appl. Phys. **61,** 3919-3921 (1987).

[69] A. Bossavit, "A rationale for 'edge-elements' in 3-D fields computations," IEEE Trans. Magn. **MAG-24,** 74-79 (1988).

[70] A. Bossavit and J.C. Vérité, "A mixed FEM-BIEM method to solve 3-D eddy-current problems," IEEE Trans. Magn. **MAG-18,** 431-435 (1982).

[71] A. Bossavit and J.C. Vérité, "The 'TRIFOU' code: Solving the 3-D eddy-currents problem by using H as state variable," IEEE Trans. Magn. **MAG-19,** 2465-2470 (1983).

[72] T. Nakata, N. Takahashi, K. Fujiwara and Y. Shiraki, "Comparison of different finite elements for 3-D eddy current analysis," IEEE Trans. Magn. **MAG-26,** 434-437 (1990).

[73] A. Kameari, "Calculation of 3D eddy current using edge-elements," IEEE Trans. Magn. **MAG-26,** 466-469 (1990).

[74] Z. Ren, F. Bouillault, A. Razek, A. Bossavit, and J.C. Vérité, "A new hybrid model using electric field formulation for 3-D eddy current problems," IEEE Trans. Magn. **MAG-26,** 470-473 (1990).

[75] A. Bossavit and I. Mayergoyz, "Edge-elements for scattering problems," IEEE Trans. Magn. **MAG-25,** 2816-2821 (1989).

[76] A. Bossavit, "Solving Maxwell equations in a closed cavity, and the question of 'spurious modes'," IEEE Trans. Magn. **MAG-26,** 702-705 (1990).

[77] J.H. Wilkinson, **The algebraic eigenvalue problem,** London: Oxford University Press, 1965.

[78] M. Hano, "Finite-element analysis of dielectric-loaded waveguides," IEEE Trans. Microwave Theory Tech. **MTT-32,** 1275-1279 (1984).

[79] C.W. Crowley, P.P. Silvester and H. Hurwitz, Jr., "Covariant projection elements for 3D vector field problems," IEEE Trans. Magn. **MAG-24,** 397-400 (1988).

[80] M. Hano, "Vector finite-element solution of anisotropic waveguides using novel triangular elements," Electron. Commun. Japan, Part 2, **71**, 71-80 (1988).

[81] A. Konrad, "A direct three-dimensional finite element method for the solution of electromagnetic fields in cavities," IEEE Trans. Magn. **MAG-21**, 2276-2279 (1985).

[82] K. Hayata, M. Koshiba, M. Eguchi and M. Suzuki, "Vectorial finite-element method without any spurious solutions for dielectric waveguiding problems using transverse magnetic-field component," IEEE Trans. Microwave Theory Tech. **MTT-34**, 1120-1124 (1986).

[83] A.J. Kobelansky and J.P. Webb, "Eliminating spurious modes in finite-element waveguide problems by using divergence-free fields," Electron. Lett. **22**, 569-570 (1986).

[84] M. Israel and R. Miniowitz, "Hermitian finite-element method for inhomogeneous waveguides," IEEE Trans. Microwave Theory Tech. **MTT-38**, 1319-1327 (1990).

[85] J.R. Winkler and J.B. Davies, "Elimination of spurious modes in finite element analysis," J. Comput. Phys. **56**, 1-14 (1984).

[86] B.M.A. Rahman and J.B. Davies, "Penalty function improvement of waveguide solution by finite elements," IEEE Trans. Microwave Theory Tech. **MTT-32**, 922-928 (1984).

[87] M. Koshiba, K. Hayata and M. Suzuki, "Improved finite-element formulation in terms of the magnetic field vector for dielectric waveguides," IEEE Trans. Microwave Theory Tech. **MTT-33**, 227-233 (1985).

[88] J.P. Webb, "The finite-element method for finding modes of dielectric-loaded cavities," IEEE Trans. Microwave Theory Tech. **MTT-33**, 635-639 (1985).

[89] B.M.A. Rahman and J.B. Davies, "Finite-element solution of integrated optical waveguides," J. Lightwave Technol. **LT-2**, 682-688 (1984).

[90] J.R. Brauer, **What every engineer should know about finite element analysis**, New York: Marcel Dekker, 1988.

[91] D.H. Norrie and G. de Vries, **The finite element method,** New York: Academic Press, 1973.

[92] T.H.H. Pian and P. Tong, "Mixed and hybrid finite-element methods" in **Finite element handbook** (ed. in chief H. Kardestuncer, project ed. D.H. Norrie), McGraw-Hill, 1987.

[93] P.G. Ciarlet, **The finite element method for elliptic problems,** Amsterdam: North-Holland Publishing Company, 1978.

[94] P.A. Raviart and J.M. Thomas, "A mixed finite element method for 2nd order elliptic problems," in **Mathematical aspects of finite element methods.** Proceedings of the Conference held in Rome, December 10-12, 1975 (Lecture notes in mathematics **606,** eds: A. Dold and B. Eckmann). Berlin Heidelberg New York: Springer-Verlag, 1977.

[95] J.C. Nedelec, "Mixed finite elements in R^3," Numer. Math. **35,** 315-341 (1980).

[96] A. Bossavit, "Whitney forms: a class of finite elements for three-dimensional computations in electromagnetism," IEE Proc., Part A, **135,** 493-500 (1988).

[97] H. Kanayama, H. Motoyama, K. Endo and F. Kikuchi, "Three-dimensional magnetostatic analysis using Nedelec's elements," IEEE Trans. Magn. **MAG-26,** 682-685 (1990).

[98] M. Ohtaka and T. Kobayashi, "A new vector variational expression for spurious-free finite-element analysis of waveguide eigenmodes," Electron. Commun. Japan Part 2, **73,** 1-8 (1990).

[99] J.T. Oden, **Finite elements of nonlinear continua,** New York: McGraw-Hill, 1972.

[100] O.C. Zienkiewicz, **The finite element method,** 3rd edn, London: McGraw-Hill, 1977.

[101] D.H. Norrie and G. de Vries, **An introduction to finite element analysis,** New York: Academic Press, 1978.

[102] P.P. Silvester and R.L. Ferrari, **Finite elements for electrical engineers,** 2nd edn, Cambridge: Cambridge University Press, 1990.

[103] A. Konrad, "Vector variational formulation of electromagnetic fields in anisotropic media," IEEE Trans. Microwave Theory Tech. **MTT-24,** 553-559 (1976).

[104] R.L. Ferrari and G.L. Maile, "Three-dimensional finite-element method for solving electromagnetic problems," Electron. Lett. **14,** 467-468 (1978).

[105] M. de Pourcq, "Field and power-density calculations by three-dimensional finite elements," IEE Proc. Part H, **130,** 377-384 (1983).

[106] R.D. Milne, **Applied functional analysis: an introductory treatment,** Boston: Pitman Advanced Publishing Program, 1980, 399-400.

[107] G. Strang and G.J. Fix, **An analysis of the finite element method,** Englewood Cliffs, N.J.: Prentice-Hall, 1973.

[108] J.R. Whiteman, "Finite-element method with singularities," in **Finite element handbook** (ed. in chief H. Kardestuncer, project ed. D.H. Norrie), McGraw-Hill, 1987.

[109] I. Babuška, "Trends in finite elements," IEEE Trans. Magn. **MAG-25,** 2799-2803 (1989).

[110] W.C. Thacker, "A brief review of techniques for generating irregular computational grids," Int. J. Numer. Methods Eng. **15,** 1335-1341 (1980).

[111] J.A. Stricklin, W.S. Ho, E.Q. Richardson, and W.E. Haisler, "On isoparametric *vs* linear strain triangular elements," Int. J. Numer. Methods Eng. **11,** 1041-1043 (1977).

[112] S.H. Lo, "A new mesh generation scheme for arbitrary planar domains," Int. J. Numer. Methods Eng. **21,** 1403-1426 (1985).

[113] J.C. Cavendish, "Automatic triangulation of arbitrary planar domains for the finite element method," Int. J. Numer. Methods Eng. **8,** 679-694 (1974).

[114] O.C. Zienkiewicz and D.V. Phillips, "An automatic mesh generation scheme for plane and curved surfaces by 'isoparametric' coordinates," Int. J. Numer. Methods Eng. **3,** 519-528 (1971).

[115] W.J. Gordon and C.A. Hall, "Construction of curvilinear co-ordinate systems and applications to mesh generation," Int. J. Numer. Methods Eng. **7**, 461-477 (1973).

[116] S.H. Lo, A.Y.T. Leung and Y.K. Cheung, "Automatic finite element mesh generation," Proc. Int. Conf. on Finite Element Method, Beijing, China, 931-937 (1982).

[117] S. Park and C.J. Washam, "Drag method as a finite element mesh generation scheme," Comp. Struct. **10**, 343-346 (1979).

[118] R.B. Haber, M.S. Shephard, J.F. Abel, R.H. Gallagher, and D.P. Greenberg, "A general two-dimensional finite element preprocessor utilizing discrete transfinite mappings," Int. J. Numer. Methods Eng. **17**, 1015-1044 (1981).

[119] R.E. Bank and A.H. Sherman, "An adaptive, multilevel method for elliptic boundary value problems," Computing **26**, 91-105 (1981).

[120] W.C. Rheinboldt, "Adaptive mesh refinement processes for finite element solutions," Int. J. Numer. Methods Eng. **17**, 649-662 (1981).

[121] R.D. Ettinger, F.A. Fernandez and J.B. Davies, "Application of adaptive remeshing techniques to the finite element analysis of nonlinear optical waveguides," **Directions in Electromagnetic Wave Modelling** [Proc. Int. Conf., eds: H.L. Bertoni and L.B. Felsen], New York: Plenum, 239-246 (1991).

[122] G.F. Carey and M. Seager, "Projection and iteration in adaptive finite element refinement," Int. J. Numer. Methods Eng. **21**, 1681-1685 (1985).

[123] J. Fukuda and J. Suhara, "Automatic mesh generation for finite element analysis," in **Advances in Computational methods in structural mechanics and design** (eds: J.T. Oden, R.W. Clough and Y. Yamamoto), Huntsville, Alabama: UAH Press, 1974, 607-624.

[124] R.D. Shaw and R.G. Pitchen, "Modification to the Suhara-Fukuda method of network generation," Int. J. Numer. Methods Eng. **12**, 93-99 (1978).

[125] A. Bykat, "Automatic generation of triangular grid: I—Subdivision of a general polygon into convex subregions. II—Triangulation of convex polygons." Int. J. Numer. Methods Eng. **10**, 1329-1342 (1976).

[126] B. Joe and R.B. Simpson, "Triangular meshes for regions of complicated shapes," Int. J. Numer. Methods Eng. **23**, 751-778 (1986).

[127] G.C. Everstine, "A comparison of three resequencing algorithms for the reduction of matrix profile and wavefront," Int. J. Numer. Methods Eng. **14**, 837-853 (1979).

[128] E.H. Cuthill and J.M. McKee, "Reducing the bandwidth of sparse symmetric matrices," Proc. 24th Nat. Conf. ACM, NJ: Barndon Systems Press, 1969, 157-172.

[129] J.W.H. Liu and A.H. Sherman, "Comparative analysis of the Cuthill-McKee and the reverse Cuthill-McKee ordering algorithms for sparse matrices," SIAM J. Numer. Anal. **13**, 197-213 (1976).

[130] N.E. Gibbs, W.G. Poole, Jr. and P.K. Stockmeyer, "An algorithm for reducing the bandwidth and profile of a sparse matrix," SIAM J. Num. Anal. **13**, 236-250 (1976).

[131] R. Levy, "Resequencing of the structural stiffness matrix to improve computational efficiency," Jet Propulsion Laboratory Quart. Tech. Review **1**, 61-70 (1971).

[132] H.R. Grooms, "Algorithm for matrix bandwidth reduction," J. Struct. Div. ASCE **98**, 203-214 (1972)

[133] J. Puttonen, "Simple and effective bandwidth reduction algorithm," Int. J. Numer. Methods Eng. **19**, 1139-1152 (1983).

[134] G. Akhras and G. Dhatt, "An automatic node relabelling scheme for minimizing a matrix or network bandwidth," Int. J. Numer. Methods Eng. **10**, 787-797 (1976).

[135] A. George and J.W.H. Liu, "A minimal storage implementation of the minimum degree algorithm," SIAM J. Numer. Anal. **17**, 282-199 (1980).

[136] I.W. Burgess and P.K.F. Lai, "A new node renumbering algorithm for bandwidth reduction," Int. J. Numer. Methods Eng. **23**, 1693-1704 (1986).

[137] I.W. Burgess and P.K.F. Lai, "Correction: A new node renumbering algorithm for bandwidth reduction," Int. J. Numer. Methods Eng. **24**, 839 (1987).

[138] I. Konishi, N. Shiraishi and T. Taniguchi, "Reducing the bandwidth of structural stiffness matrices," J. Struct. Mech. **4**, 197-226 (1976).

[139] R.J. Collins, "Bandwidth reduction by automatic renumbering," Int. J. Numer. Methods Eng. **6**, 345-356 (1973).

[140] N.E. Gibbs, W.G. Poole, Jr. and P.K. Stockmeyer, "A comparison of several bandwidth and profile reduction algorithms," ACM Trans. on Math. Soft. **2**, 322-330 (1976).

[141] P. Rabinowitz (ed.), **Numerical methods for nonlinear algebraic equations,** London: Gordon and Breach Science Publishers, 1970.

[142] J.C.P. Bus, **Numerical solution of systems of nonlinear equations,** Amsterdam: Mathematisch Centrum, 1980.

[143] M. Geradin and M. Hogge, "Solving systems of nonlinear equations," in **Finite element handbook** (ed. in chief H. Kardestuncer, project ed. D.H. Norrie), New York: McGraw-Hill, 1987.

[144] M.A. Wolfe, **Numerical methods for unconstrained optimization − an introduction,** New York: Van Nostrand Rheinhold, 1978.

[145] M. Geradin, S. Idelsohn and M. M. Hogge, " Computational strategies for the solution of large nonlinear problems *via* quasi-Newton methods," Comput. Struct. **13,** 73-81 (1981).

[146] B.M. Irons and A. Elsawaf, "The conjugate Newton algorithm for solving finite element equations," in **Proceedings U.S. German Symposium on formulations and algorithms in finite element analysis** (eds: K.J. Bathe, J.T. Oden and W. Wunderlich), Cambridge: Massachusetts Institute of Technology, Massachusetts, 1977, 656-672.

[147] J.E. Dennis and J.J. More, "Quasi-Newton methods, motivation and theory," SIAM Rev. **19**, 46-89 (1977).

[148] H. Mathies and G. Strang, "The solution of nonlinear finite element equations," Int. J. Numer. Methods Eng. **14,** 1613-1626 (1979).

[149] K.J. Bathe and A. Cimento, "Some practical procedures for the solution of nonlinear finite element equations," Comput. Meth. Appl. Mech. Eng. **22,** 59-85 (1980).

[150] L.K. Schubert, "Modification of a quasi-Newton method for nonlinear equations with with a sparse Jacobian," Math. Comp. **24,** 27-30 (1970).

[151] M.A. Crisfield, "A faster modified Newton-Raphson iteration," Comput. Meth. Appl. Mech. Eng. **20,** 267-278 (1979).

[152] M. Papadrakakis and C.J. Gantes, "Preconditioned conjugate- and secant-Newton methods for non-linear problems," Int. J. Numer. Methods Eng. **28,** 1299-1316 (1989).

[153] R.D. Ettinger, F.A. Fernandez, B.M.A. Rahman and J.B. Davies, "Vector finite element solution of saturable nonlinear strip-loaded optical waveguides," IEEE Photon. Technol. Lett. **3,** 147-149 (1991).

[154] A. Palazzolo and W.D. Pilkey, "Eigensolution extraction methods" in **Finite element handbook** (ed. in chief H. Kardestuncer, project ed. D.H. Norrie), New York: McGraw-Hill, 1987.

[155] A. Jennings, **Matrix computation for engineers and scientists,** New York: John Wiley & Sons, 1977.

[156] L.A. Hageman and D.M. Young, **Applied iterative methods,** New York: Academic Press, 1981.

[157] H.R. Schwarz, H. Rutishauser and E. Stiefel, **Numerical analysis of symmetric matrices,** Englewood Cliffs, N.J.: Prentice-Hall, 1973.

[158] G. Peters and J.H. Wilkinson, "$Ax = \lambda Bx$ and the generalized eigenproblem," SIAM J. Numer. Anal. **7,** 479-492 (1970).

[159] K.K. Gupta, "Eigenproblem solution by a combined Sturm sequence and inverse iteration technique," Int. J. Numer. Methods Eng. **7,** 17-42 (1973).

[160] M. Geradin, "Solution algorithms for static and eigenvalue problems" in **Finite element handbook** (ed. in chief H. Kardestuncer, project ed. D.H. Norrie), New York: McGraw-Hill, 1987.

[161] G.F. Carey and J.T. Oden, **Finite elements: computational aspects,** Vol. III Englewood Cliffs, N.J.: Prentice-Hall, 1984.

[162] D.H. Sinnott, "The computation of waveguide fields and cut-off frequencies using finite difference techniques," Weapons Research Establishment, Salisbury, South Australia, Tech. Note PAD 158 (1969).

[163] D.H. Sinnott, G.K. Cambrell, C.T. Carson and H.E. Green, "The finite difference solution of microwave circuit problems," IEEE Trans. Microwave Theory Tech. **MTT-17,** 464-478 (1969).

[164] C.G. Williams, **Finite element analysis of closed and open-boundary inhomogeneous waveguides,** Ph.D. Thesis, Monash University, Melbourne, Australia, 1976.

[165] C.B. Moler, "Finite difference methods for the eigenvalues of Laplace's operator," Computer Science Department, Stanford University, Stanford, Calif., Tech. Rept. CS 22 (1965).

[166] D.S. Kershaw, "The incomplete Choleski-conjugate gradient method for the iterative solution of systems of linear equations," J. Comp. Phys. **26,** 43-65 (1978).

[167] M.A. Ajiz and A. Jennings, "A robust incomplete Choleski-conjugate gradient algorithm," Int. J. Numer. Methods Eng. **20,** 949-966 (1984).

[168] M. Papadrakakis, "Solution of the partial eigenproblem by iterative methods," Int. J. Numer. Methods Eng. **20,** 2283-2301 (1984).

[169] M. Papadrakakis, "Accelerating vector iteration methods," J. Appl. Mech. **53,** 291-297 (1986).

[170] A.K. Noor and J.M. Peters, "Preconditioned conjugate gradient technique for the analysis of symmetric anisotropic structures," Int. J. Numer. Methods Eng. **24,** 2057-2070 (1987).

[171] M. Papadrakakis and M. Yakoumidakis, "On the preconditioned conjugate gradient method for solving $(A - \lambda B)X = 0$," Int. J. Numer. Methods Eng. **24,** 1355-1366 (1987).

[172] B. Engquist, A. Greenbaum and W.D. Murphy, "Global boundary conditions and fast Helmholtz solvers," IEEE Trans. Magn. **MAG-25**, 2804-2806 (1989).

[173] G. Peters and J.H. Wilkinson, "Eigenvalues of $Ax = \lambda Bx$ with band symmetric A and B," Comput. J. **12**, 398-404 (1969).

[174] G.S. Whiston, "On an eigensystem routine due to K.K. Gupta," Int. J. Numer. Methods Eng. **10**, 759-764 (1976).

[175] A.J. Jennings, **Refined finite element analysis of anisotropic acoustic waveguides**, Ph.D. thesis, Monash University, Melbourne, Australia, 1979.

[176] F.W. Williams and D. Kennedy, "Reliable use of determinants to solve nonlinear structural eigenvalue problems efficiently," Int. J. Numer. Methods Eng. **26**, 1825-1841 (1988).

[177] R.E. Collin, **Field theory of guided waves,** 2nd edn, IEEE Press, 1991.

[178] D.T. Thomas, "Functional approximations for solving boundary value problems by computer," IEEE Trans. Microwave Theory Tech. **MTT-17**, 447-484 (1969).

[179] M. Koshiba, K. Hayata and M. Suzuki, "Vector finite-element formulation without spurious modes for dielectric waveguides," Trans. IECE Japan **E 67**, 191-196 (1984).

[180] F.A. Fernandez and Y. Lu, "Variational finite element analysis of dielectric waveguides with no spurious solutions," Electron. Lett. **26**, 2125-2126 (1990).

[181] A.C. Aitkin, "On the iterative solution of a system of linear equations," Proc. Roy. Soc. Edinburgh **63**, 52-60 (1950).

[182] O.C. Zienkiewicz and B.M. Irons, "Matrix iteration and acceleration process in finite element problems of structural mechanics" in **Numerical methods for nonlinear algebraic equations** (ed. P. Rabinowitz), London: Gordon and Breach, Science Publishers Ltd., 1970, 183-194.

[183] X.H. Wang, G.K. Cambrell and L.N. Binh, "A package for nonlinear optical waveguides based on *E*-vector finite elements," in **Advances in electrical engineering software** [Proceedings of the First International Conference on Electrical Engineering Analysis and Design, Lowell, Massachusetts, USA, 21-23 August 1990] (ed. P.P. Silvester), Southampton Boston: Computational Mechanics Publications, 1990, 151-162.

[184] J.M. Arnold, "Nonlinear guided waves," Radio Science **28**, 885-890 (1993).

[185] J.E. Ehrlich, G. Assanto and G.I. Stegeman, "Nonlinear guided-wave grating phenomena," Proceedings of the SPIE **1280**, 136-149 (1990).

[186] A.D. Boardman and T. Twardowski, "Transverse-electric and transverse-magnetic waves in nonlinear isotropic waveguides," Phys. Rev. A **39**, 2481-2492 (1989).

[187] M. Haelterman, "Pure phase bistability with a nonlinear slab waveguide," Opt. Lett. **13**, 791-793 (1988).

[188] X.H. Wang, L.N. Binh and G.K. Cambrell, "Vectorial finite-element methods for nonlinear optical waveguides," Proc. 13th Australian Conference on Optical Fibre Technology, IREE Aust., Hobart, Australia, December 1988, 129-132.

[189] X.H. Wang, G.K. Cambrell and L.N. Binh, "Scalar and vector formulations of nonlinear optical waveguides: a comparison," Proc. IREECON International 1989, IREE Aust., Melbourne, Australia, September 1989, 551-554.

[190] N.N. Akhmediev, R.F. Nabiev and Yu.M. Popov, "Stripe nonlinear surface waves," Solid State Commun. **66**, 981-985 (1988).

[191] N.N. Akhmediev, R.F. Nabiev and Yu.M. Popov, "Three-dimensional modes of a symmetric nonlinear plane waveguide," Opt. Commun. **69**, 247-252 (1989).

[192] N.N. Akhmediev, R.F. Nabiev and Yu.M. Popov, "Stripe nonlinear waves in a symmetrical planar structure," Opt. Commun. **72**, 190-194 (1989).

[193] X.H. Wang and G.K. Cambrell, "All-optical switching and bistability phenomena in nonlinear optical waveguides: Part I Power
dispersion relations," Proc. 16th Australian Conference on Optical Fibre Technology, IREE Aust., Adelaide, Australia, December
1991, 314-317.

[194] X.H. Wang and G.K. Cambrell, "Full vectorial simulation of bistability phenomena in nonlinear optical waveguides," J. Opt. Soc.
Am. B, **10**, 1090-1095 (1993).

[195] D.J. Robbins, "TE modes in a slab waveguide bounded by nonlinear media," Opt. Commun. **47**, 309-312 (1983).

[196] U. Langbein, F. Lederer, H.-E. Ponath and U. Trutschel, "Dispersion relations for nonlinear guided waves," J. Molecular Struct.
115, 493-496 (1984).

[197] C.T. Seaton, J.D. Valera, B. Svenson and G.I. Stegeman, "Comparison of solutions for TM-polarized nonlinear guided waves," Opt.
Lett. **10**, 149-150 (1985).

[198] U. Langbein, F. Lederer and H.-E. Ponath, "Generalized dispersion
relations for nonlinear slab-guided waves," Opt. Commun. **53**, 417-
420 (1985).

[199] A.D. Boardman and P. Egan, "S-polarized waves in a thin dielectric
film asymmetrically bounded by optically nonlinear media," IEEE
J. Quantum Electron. **QE-21**, 1701-1713 (1985).

[200] C.T. Seaton *et al.,* "Calculation of nonlinear TE waves guided by
thin dielectric films bounded by nonlinear media," IEEE J. Quantum Electron. **QE-21**, 774-783 (1985).

[201] A.D. Boardman and P. Egan, "Optically nonlinear waves in thin
films," IEEE J. Quantum Electron. **QE-22**, 319-324 (1986).

[202] R.K. Varshney, M.A. Nehme, R. Srivastava and R.V. Ramaswamy,
"Guided waves in graded-index planar waveguides with nonlinear
cover medium," Appl. Opt. **25**, 3899-3902 (1986).

[203] W.R. Holland, "Nonlinear guided waves in low-index, self-focussing
thin films: transverse electric case," J. Opt. Soc. Am. B **3**, 1529-
1534 (1986).

[204] D. Milhalache *et al.,* "Exact dispersion relations for transverse magnetic polarized guided waves at a nonlinear interface," Opt. Lett. **12,** 187-189 (1987).

[205] K. Ogusu, "TE waves in a symmetric dielectric slab waveguide with a Kerr-like nonlinear permittivity," Opt. Quantum Electron. **19,** 65-72 (1987).

[206] S. Chelkowski and J. Chrostowski, "Scaling rules for slab waveguides with nonlinear substrate," Appl. Opt. **26,** 3681-3686 (1987).

[207] T. Sakakibara and N. Okamoto, "Nonlinear TE waves in a dielectric slab waveguide with two optically nonlinear layers," IEEE J. Quantum Electron. **QE-23,** 2084-2088 (1987).

[208] K. Ogusu, "Computer analysis of general nonlinear planar waveguides," Opt. Commun. **64,** 425-430 (1987).

[209] S.Y. Shin, E.M. Wright and G.I. Stegeman, "Nonlinear TE waves of coupled waveguides bounded by nonlinear media," J. Lightwave Technol. **6,** 977-983 (1988).

[210] U. Trutschel, F. Lederer and M. Golz, "Nonlinear guided waves in multilayer systems," IEEE J. Quantum Electron. **25,** 194-200 (1989).

[211] K. Ogusu, "Analysis of non-linear multilayer waveguides with Kerr-like permittivities," Opt. Quantum Electron. **21,** 109-116 (1989).

[212] A. Heinämäki and S. Honkanen, "Non-linear waves in a multilayer waveguide: application to SiO_xN_y with organic overlayer," Opt. Quantum Electron. **21,** 137-146 (1989).

[213] S.J. Al-Bader and H.A. Jamid, "Guided waves in nonlinear saturable self-focusing thin films," IEEE J. Quantum Electron. **QE-23,** 1947-1955 (1987).

[214] S. Vukovic and R. Dragila, "Power-flux-dependent polarization of nonlinear surface waves," Opt. Lett. **15,** 168-170 (1990).

[215] N. Saiga, "Calculation of TE modes in graded-index nonlinear optical waveguides with arbitrary profile of refractive index," J. Opt. Soc. Am. B **8,** 88-94 (1991).

[216] R.A. Sammut and C. Pask, "Gaussian and equivalent-step-index approximations for nonlinear waveguides," J. Opt. Soc. Am. B **8**, 395-402 (1991).

[217] D. Marcuse, "Reflection of a Gaussian beam from a nonlinear interface," Appl. Opt. **19**, 3130-3139 (1980).

[218] W.J. Tomlinson, J.P. Gordon, P.W. Smith and A.E. Kaplan, "Reflection of a Gaussian beam at a nonlinear interface," Appl. Opt. **21**, 2041-2051 (1982).

[219] K. Hayata, M. Koshiba and M. Suzuki, "Finite-element solution of arbitrary nonlinear, graded-index slab waveguides," Electron. Lett. **23**, 429-431 (1987).

[220] A.D. Boardman, T. Twardowski and E.M. Wright, "The effect of diffusion on surface-guided nonlinear TM waves: a finite element approach," Opt. Commun. **74**, 347-352 (1990).

[221] A.D. McAulay, X. Xu and J. Wang, "Designing fast optically controlled waveguide switches," Proceedings of the SPIE **2026**, 268-275 (1993).

[222] M. Zoboli and S. Selleri, "Finite element analysis of TE and TM modes in nonlinear planar waveguides," Int. J. Nonlinear Opt. Phys. **3**, 101-116 (1994).

[223] A.D. Boardman, K. Singh, J. Hoad and T. Twardowski, "Linear and nonlinear rib waveguides with channel tapering," Opt. Quantum Electron. **26**, S321-S334 (1994).

[224] S.R. Cvetkovic, A.P. Zhao and M. Punjani, "Automated finite-element solution of nonlinear optical waveguide problems in two dimensions," Microwave Opt. Technol. Lett. **7**, 293-296 (1994).

[225] S.R. Cvetkovic and A.P. Zhao, "Finite-element formalism for linear and nonlinear guided waves in multiple-quantum-well waveguides," J. Opt. Soc. Am. B **10**, 1401-1407 (1993).

[226] A.P. Zhao and S.R. Cvetkovic, "Finite-element solution of nonlinear TM waves in multiple-quantum-well waveguides," IEEE Photon. Technol. Lett. **4**, 1231-1234 (1992).

[227] Q.Y. Li, R.A. Sammut and C. Pask, "Variational and finite element analyses of nonlinear strip optical waveguides," Opt. Commun. **94**, 37-43 (1992).

[228] A.P. Zhao and R.S. Cvetkovic, "Numerical modeling of nonlinear TE waves in multiple-quantum-well waveguides," IEEE Photon. Technol. Lett. **4**, 623-626 (1992).

[229] H.E. Hernandez-Figueroa, "A new finite element scheme for optical temporal soliton analysis," Int. Conf. on Computation in Electromagnetics, London: IEE, November 1991, 167-170.

[230] K. Hayata and M. Koshiba, "Mutual guiding assistance between eigenmodes of nonlinearly coupled TE-TM waves," Trans. Institute Electron. Information Commun. Engineers E **E74**, 2890-2897 (1991).

[231] M. Zoboli, F. Di Pasquale and P. Bassi, "Analysis of nonlinear bistable optical waveguides by a full vectorial finite-element method," Proceedings of the European Conference on Optics, Optical Systems and Applications (ECOOSA 90 - Quantum Optics, Rome, Italy, November 1990), Bristol: IOP, 1991, 157-160.

[232] X. H. Wang and G. K. Cambrell, "Simulation of strong nonlinear efffects in optical waveguides," J. Opt. Soc. Am. B, **10**, 2048-2055 (1993)

[233] K. Hayata and M. Koshiba, "Self-localization and spontaneous symmetry breaking of optical fields propagating in strongly nonlinear channel waveguides: limitations of the scalar field approximation," J. Opt. Soc. Am. B, **9**, 1362-1368 (1992).

[234] X.H. Wang, L.N. Binh and G.K. Cambrell, "Numerical analysis of a nonlinear optical channel waveguide," Proc. 14th Australian Conference on Optical Fibre Technology, IREE Aust., Brisbane, Australia, December 1989, 225-228.

[235] R. Cuykendall and K.H. Strobl, "Effects of soft saturation on nonlinear interface switching," Phys. Rev. A **41**, 352-358 (1990).

[236] W.C. Banyai *et al.*, "Saturation of the nonlinear refractive-index change in a semiconductor-doped glass channel waveguide," Appl. Phys. Lett. **54**, 481-483 (1989).

[237] S.Y. Auyang and P.A. Wolff, "Free-carrier-induced third-order optical nonlinearities in semiconductors," J. Opt. Soc. Am. B **6**, 595-605 (1989).

[238] J.L. Coutaz and M. Kull, "Saturation of the nonlinear index of refraction in semiconductor-doped glass," J. Opt. Soc. Am. B **8**, 95-98 (1991).

[239] Y. Silberberg, "Solitons and two-photon absorption," Opt. Lett. **15**, 1005-1007 (1990).

[240] K. Hayata, A. Misawa and M. Koshiba, "Finite-element simulation of nonlinear optical guided-wave propagation using direct integration scheme," Electron. Commun. Japan, Part 2, **74**, 13-20 (1991).

[241] V.P. Nayyar, A. Kumar and D. Kaur, "Non-linear propagation of a mixture of two elliptically polarised degenerate modes of a laser beam," Opt. Commun. **79**, 93-98 (1990).

[242] K. Ogusu, "Nonlinear wave propagation along a bent nonlinear dielectric interface," Opt. Lett. **16**, 312-314 (1991).

[243] K. Ogusu, "Stability of a new type of stationary waves guided by a nonlinear hollow waveguide," Opt. Commun. **83**, 260-264 (1991).

[244] X. H. Wang and G. K. Cambrell, "Vectorial simulation and power-parameter characterisation of nonlinear planar optical waveguides," J. Opt. Soc. Am. B **12**, February 1995.

[245] S.M. Jensen, "The nonlinear coherent coupler," IEEE J. Quantum Electron. **QE-18**, 1580-1583 (1982).

[246] G.I. Stegeman, C.T. Seaton, C.N. Ironside, T. Cullen, and A.C. Walker, "Effects of saturation and loss on nonlinear directional couplers," Appl. Phys. Lett. **50**, 1035-1037 (1987).

[247] L.-P. Yuan, "A unified approach for the coupled-mode analysis of nonlinear optical couplers," IEEE J. Quantum Electron. **30**, 126-133 (1994).

[248] F.J. Fraile-Pelaez and G. Assanto, "Improved coupled-mode analysis of nonlinear distributed feedback structures," Opt. Quantum Electron. **23**, 633-637 (1991).

[249] E. Weinert-Raczka, "Nonlinear TE-TM mode coupling in a thin-film optical waveguide," Proceediogs of the SPIE **859**, 216-219 (1988).

[250] E. Caglioti, S. Trillo, S. Wabnitz, B. Daino and G.I. Stegeman "Power-dependent switching in a coherent nonlinear directional coupler in the presence of saturation," Appl. Phys. Lett. **51**, 293-295 (1987).

[251] G.I. Stegeman, E. Caglioti, S. Trillo and S. Wabnitz, "Parameter trade-offs in nonlinear directional couplers: two level saturable nonlinear media," Opt. Commun. **63**, 281-284 (1987).

[252] A.W. Snyder, Y. Chen and A. Ankiewicz, "Coupled waves on optical fibers by power conservation," J. Lightwave Technol. **7**, 1400-1406 (1989).

[253] C. Vassilopoulos and J.R. Cozens, "Combined directional and contradirectional coupling in a three-waveguide configuration," IEEE J. Quantum Electron. **25**, 2113-2118 (1989).

[254] Y. Chen, "Solution to full coupled wave equations of nonlinear coupled systems," IEEE J. Quantum Electron. **25**, 2149-2153 (1989).

[255] E.M. Wright, D.R. Healey, G.I. Stegeman and K.J. Blow, "Variation of the switching power with diffusion length in a nonlinear directional coupler," Opt. Commun. **73**, 385-392 (1989).

[256] F.J. Fraile-Pelaez, G. Assanto and D.R. Heatley, "Sign-dependent response of nonlinear directional couplers," Opt. Commun. **77**, 402-406 (1990).

[257] D.J. Mitchell, A.W. Snyder and Y. Chen, "Nonlinear triple core couplers," Electron. Lett. **26**, 1164-1166 (1990).

[258] C. Schmidt-Hattenberger, U. Trutschel and F. Lederer, "Nonlinear switching in multiple-core couplers," Opt. Lett. **16**, 294-296 (1991).

[259] M.D. Feit and J.A. Fleck, Jr., "Three-dimensional analysis of a directional coupler exhibiting a Kerr nonlinearity," IEEE J. Quantum Electron. **24**, 2081-2086 (1988).

[260] N. Finlayson, E.M. Wright, C.T. Seaton, G.I. Stegeman, and Y. Silberberg, "Beam propagation study of nonlinear coupling between transverse electric modes of a slab waveguide," Appl. Phys. Lett. **50**, 1562-1564 (1987).

[261] H.E. Hernandez-Figueroa, "Improved split-step schemes for nonlinear-optical propagation," J. Opt. Soc. Am. B **11**, 798-803 (1994).

[262] D. Mihalache and D. Mazilu, "Propagation phenomena of nonlinear guided waves in graded-index planar waveguides," IEE Proc. J: Optoelectron. **138**, 365-372 (1991).

[263] K. Hayata, A. Misawa and M. Koshiba, "Chaotic propagation of the elliptically-polarized light in nonlinear guided-wave structures," Trans. Institute Electron. Information Commun. Engineers E **E73**, 855-860 (1990).

[264] K. Hayata and M. Koshiba, "Bistability analysis of nonlinear optical guided waves," Electron. Commun. Jpn, Part 2, **74**, 21-30 (1991).

[265] K. Hayata, A. Misawa and M. Koshiba, "Nonstationary simulation of nonlinearly coupled TE-TM waves propagating down dielectric slab structures by the step-by-step finite-element method," Opt. Lett. **15**, 24-26 (1990).

[266] K. Hayata, A. Misara and M. Koshiba, "Split-step finite-element method applied to nonlinear integrated optics," J. Opt. Soc. Am. B **7**, 1772-1784 (1990).

[267] J.V. Moloney, J. Ariyasu, C.T. Seaton and G.I. Stegeman, "Stability of nonlinear stationary waves guided by a thin film bounded by nonlinear media," Appl. Phys. Lett. **48**, 826-828 (1986).

[268] J. Ariyasu, C.T. Seaton, G.I. Stegeman and J.V. Moloney, "New theoretical developments in nonlinear guided waves: stability of TE_1 branches," IEEE J. Quantum Electron. **QE-22**, 984-987 (1986).

[269] R.A. Sammut, Q.Y. Li and C. Pask, "Variational approximations and mode stability in planar nonlinear waveguides," J. Opt. Soc. Am. B **9**, 884-890 (1992).

[270] D. Mihalache and D. Mazilu, "Stability of nonlinear stationary slab-guided waves in saturable media: a numerical analysis," Phys. Lett. A **122**, 381-384 (1987).

[271] T.B. Koch, J.B. Davies and D. Wickramasinghe, "Finite element/finite difference propagation algorithm for integrated optical device," Electron. Lett. **25**, 514-516 (1989).

[272] M.D. Feit and J.A. Fleck, Jr., "Light propagation in graded-index optical fibers," Appl. Opt. **17**, 3990-3998 (1978).

[273] J. van Roey, J. van der Donk and P.E. Lagasse, "Beam-propagation method: analysis and assessment," J. Opt. Soc. Am. **71**, 803-810 (1981).

[274] J.A. Fleck, Jr. and M.D. Feit, "Beam propagation in uniaxial anisotropic media," J. Opt. Soc. Am. **73**, 920-926 (1983).

[275] D.J. Thomson and N.R. Chapman, "A wide-angle split-step algorithm for the parabolic equation," J. Acoust. Soc. Am. **74**, 1848-1854 (1983).

[276] L.R. Gomma, "Beam propagation method applied to a step discontinuity in dielectric planar waveguides," IEEE Trans. Microwave Theory Tech. **36**, 791-792 (1988).

[277] J.J. Gribble and J.M. Arnold, "Beam propagation method and geometrical optics," IEE Proc., Part J, **135**, 343-348 (1988).

[278] M.D. Feit and J.A. Fleck, Jr, "Beam nonparaxiality, filament formation and beam breakup in the self-focussing of optical beams," J. Opt. Soc. Am. B **5**, 633-640 (1988).

[279] Y. Chung and N. Dagli, "An assessment of finite difference beam propagation method," IEEE J. Quantum Electron. **26**, 1335-1339 (1990).

[280] J.A. Fleck, Jr., "A cubic spline method for solving the wave equation of nonlinear optics," J. Comput. Phys. **16**, 324-341 (1974).

[281] A. Korpel, K.E. Lonngren, P.P. Banerjee, H.K. Sim, and M.R. Chatterjee, "Split-step-type angular plane-wave spectrum method for the study of self-refractive effects in nonlinear wave propagation," J. Opt. Soc. Am. B **3**, 885-890 (1986).

[282] L. Leine, C. Wächter, U. Langbein and F. Lederer, "Propagation phenomena of nonlinear film-guided waves: a numerical analysis," Opt. Lett. **11**, 590-592 (1986).

[283] N.C. Kothari and C. Flytzanis, "Light propagation in a two-component nonlinear composite medium," Opt. Lett. **12**, 492-494 (1987).

[284] C.S. Mitchell and J.V. Moloney, "Propagation characteristics of optical beams in planar thin-film waveguides," Opt. Commun. **69**, 243-246 (1989).

[285] K. Hayata, A. Misawa and M. Koshiba, "Nonlinear beam propagation in tapered waveguides," Electron. Lett. **25**, 661-662 (1989).

[286] E.M. Wright, G.I. Stegeman and S.W. Koch, "Numerical simulation of guided-wave phenomena in semiconductors," J. Opt. Soc. Am. B **6**, 1598-1606 (1989).

[287] A.M. Kamchatnov, "Propagation of ultrashort periodic pulses in nonlinear fiber waveguides," Sov. Phys. JETP **70**, 80-84 (1990).

[288] H. Fouckhardt and Y. Silberberg, "All-optical switching in waveguide X junctions," J. Opt. Soc. Am. B **7**, 803-809 (1990).

[289] J.A. Giannini and R.I. Joseph, "The propagation of bright and dark solitons in lossy optical fibers," IEEE J. Quantum Electron. **26**, 2109-2114 (1990).

[290] C.S. Mitchell and J.V. Moloney, "Propagation and stability of optical pulses in nonlinear planar structures with instantaneous and finite response times," IEEE J. Quantum Electron. **26**, 2115-2129 (1990).

[291] N.N. Akhmediev, V.I. Korneev and Yu.V. Kuz'menko, "Stability of nonlinear waves," Sov. Tech. Phys. Lett. **10**, 327-329 (1984).

[292] J.V. Moloney, J. Ariyasu, C.T. Seaton and G.I. Stegeman, "Numerical evidence for nonstationary, nonlinear, slab-guided waves," Opt. Lett. **11**, 315-317 (1986).

[293] L. Leine, C. Wächter, U. Langbein and F. Lederer, "Evolution of nonlinear guided optical fields down a dielectric film with a nonlinear cladding," J. Opt. Soc. Am. B **5**, 547-558 (1988).

[294] D. Mihalache, D. Mazilu, M. Bertolotti and C. Sibilia, "Exact solution for nonlinear thin-film guided waves in higher-order nonlinear media," J. Opt. Soc. Am. B **5,** 565-570 (1988).

[295] B.D. Robert and J.E. Sipe, "Transverse instability in a thin nonlinear slab," Phys. Rev. A **38,** 5217-5226 (1988).

[296] W.J. Firth and C. Paré, "Transverse modulational instabilities for counterpropagating beams in Kerr media," Opt. Lett. **13,** 1096-1098 (1988).

[297] X.H. Wang and G.K. Cambrell, "All-optical switching and bistability phenomena in nonlinear optical waveguides: Part II Propagation properties," Proc. 16th Australian Conference on Optical Fibre Technology, IREE Aust., Adelaide, Australia, December 1991, 318-321.

[298] A.W. Snyder, S.J. Hewlett and D.J. Mitchell, "Dynamic spatial solitons," Phys. Rev. Lett. **72,** 1012-1015 (1994).

[299] W. Krolikowski, X. Yang, B. Luther-Davies and J. Breslin, "Dark soliton steering in a saturable nonlinear medium," Opt. Commun. **105,** 219-225 (1994).

[300] Y.S. Kivshar, "Dark solitons in nonlinear optics," IEEE J. Quantum Electron. **29,** 250-264 (1993).

[301] Y. Chen and J. Atai, "Gray and dark modes in nonlinear waveguides," J. Opt. Soc. Am. B **9,** 2252-2257 (1992).

[302] D.J. Mitchell and A.W. Snyder, "Stability of fundamental nonlinear guided waves," J. Opt. Soc. Am. B **10,** 1572-1580 (1993).

[303] J.V. Moloney, A.C. Newell and A.B. Aceves, "Spatial soliton optical switching: a soliton-based equivalent particle approach," Opt. Quantum Electron. **24,** S1269-S1293 (1992).

[304] A.W. Snyder, D.J. Mitchell and L. Poladian, "Linear approach for approximating spatial solitons and nonlinear guided modes," J. Opt. Soc. Am. B **8,** 1618-1620 (1991).

[305] S. Lakshmanasamy, A.K. Jordan and S.S. Mitra, "Soliton propagation in optical waveguides: a review," Proceedings of the XVIth Workshop on Interdisciplinary Study of Inverse Problems: Some

Topics on Inverse Problems, Singapore: World Scientific, 391-403 (1988).

[306] P.R. Berger, P.K. Bhattacharya and S. Gupta, "A waveguide directional coupler with a nonlinear coupling medium," IEEE J. Quantum Electron. **27**, 788-795 (1991).

[307] R. Horak, M. Bertolotti, C. Sibilia and J. Perina, "Quantum effects in a nonlinear coherent coupler," J. Opt. Soc. Am. B **6**, 199-204 (1989).

[308] B.P. Nelson and D. Wood, "Analysis of the nonlinear coaxial coupler," IEEE J. Quantum Electron. **24**, 1915-1921 (1988).

[309] H.C. Hsieh and P.N. Robson, "Nonlinear optical waveguide directional coupler employing multiple quantum well structure," J. Appl. Phys. **64**, 1696-1703 (1988).

[310] C. Vassallo, **Optical waveguide concepts**, New York: Elsevier Science Publishers B.V., 1991.

[311] J.R. Pierce, "Coupling of modes of propagation," J. Appl. Phys., **25**, 179-183 (1954).

[312] H.A. Haus, W.P. Huang, S. Kawakami and N.A. Whitaker, "Coupled-mode theory of optical waveguides," J. Lightwave Technol., **LT-5**, 16-23 (1987).

[313] E.A.J. Marcatili, "Dielectric rectangular waveguide and directional coupler for integrated optics," Bell Syst. Tech. J., **48**, 2071-2102 (1969).

[314] D. Marcuse, "The coupling of degenerate modes in two parallel dielectric waveguides," Bell Syst. Tech. J., **50**, 1791-1816 (1971).

[315] A.W. Snyder, "Coupled-mode theory for optical fibres," J. Opt. Soc. Am., **62**, 1267-1277 (1972).

[316] A. Yariv, "Coupled-mode theory for guided-wave optics," IEEE J. Quantum Electron., **QE-9**, 919-933 (1973).

[317] A. Hardy and W. Streifer, "Coupled-mode theory of parallel waveguides," J. Lightwave Technol., **LT-3**, 1135-1146 (1985).

[318] A. Hardy and W. Streifer, "Coupled-modes of multiwaveguides systems and phased arrays," J. Lightwave Technol., **LT-4**, 90-99 (1986).

[319] A. Hardy, S. Shakir and W. Streifer, "Coupled-mode equations for two weakly guiding single-mode fibers," Opt. Lett., **11**, 324-326 (1986).

[320] W. Streifer, M. Osinski and A. Hardy, "Reformulation of the coupled-mode theory of multiwaveguide systems", J. Lightwave Technol. **LT-5**, 1-4 (1987).

[321] J.R. Qian, "Generalized coupled-mode equations and their applications to fibre couplers," Electron. Lett., **20**, 304-306 (1986).

[322] E. Marcatili, "Improved coupled-mode equations for dielectric waveguides," IEEE J. Quantum Electron., **QE-22**, 988-993 (1986).

[323] S.L. Chuang, "A coupled-mode formulation by reciprocity and a variational principle," J. Lightwave Technol., **LT-5**, 5-15 (1987).

[324] S.L. Chuang, "Application of the strongly coupled-mode theory to integrated optical devices," IEEE J. Quantum Electron., **QE-23**, 499-509 (1987).

[325] C. Vassallo, "About coupled-mode theories for dielectric waveguides," J. Lightwave Technol., **6**, 294-303 (1988).

[326] H.S. Huang and H.C. Chang, "Vector coupled-mode analysis of coupling between two identical optical fiber cores," Opt. Lett., **14**, 90-92 (1989).

[327] H.A. Haus, W.P. Huang and A.W. Snyder, "Coupled-mode formulations," Opt. Lett., **14**, 1222-1224 (1989).

[328] W.P. Huang and H.A. Haus, "Self-consistent vector coupled-mode theory for tapered optical waveguides," J. Lightwave Technol., **8**, 922-926 (1990).

[329] W.P. Huang and S.K. Chaudhuri, "Variational coupled-mode theory of optical couplers," J. Lightwave Technol., **8**, 1565-1570 (1990).

[330] H.S. Huang and H.C. Chang, "Vector coupled-mode calculation of guided vector modes on an equilateral three-core optical fiber," IEEE Microwave Guided Wave Lett., **1**, 57-59 (1991).

[331] X.J. Meng and N. Okamoto, "Improved coupled-mode theory for nonlinear directional couplers," IEEE J. Quantum Electron., **27**, 1175-1181 (1991).

[332] W.P. Huang and J. Hong, "A coupled-mode analysis of modulation instability in optical fibers," J. Lightwave Technol., **10**, 156-162 (1992).

[333] N. Finlayson and G.I. Stegeman, "Spatial switching, instabilities and chaos in a three-waveguide nonlinear directional coupler," Appl. Phys. Lett., **56**, 2276-2278 (1990).

[334] F.J. Fraile-Pelaez and G. Assanto, "Coupled-mode equations for nonlinear directional couplers," Appl. Opt., **29**, 2216-2217 (1990).

[335] R.J. Black, A. Henault, S. Lacroix and M. Cada, "Structural considerations for bimodal nonlinear optical switching," IEEE J. Quantum Electron., **26**, 1081-1088 (1990).

[336] Y. Chen and A.W. Snyder, "Normal modes of twin core periodic structures," J. Opt. Soc. Am. B, **8**, 1621-1625 (1991).

[337] D.G. Hall, "Coupled-mode theory for corrugated optical waveguides," Opt. Lett., **15**, 619-621 (1990).

[338] C.M. de Sterke, K. R. Jackson and B. D. Robert, "Nonlinear coupled-mode equations on a finite interval: a numerical procedure," J. Opt. Soc. Am. B, **8**, 403-412 (1991).

[339] R. Schiek, "Time-resolved switching characteristic of the nonlinear directional coupler under consideration of susceptibility dispersion," IEEE J. Quantum Electron., **27**, 2150-2158 (1991).

[340] W.P. Huang and C.L. Xu, "Simulation of three-dimensional optical waveguides by a full-vector beam propagation method," IEEE J. Quantum Electron., **29**, 2639-2649 (1993).

[341] A. Splett, M. Majd and K. Petermann, "A novel beam propagation method for large refractive index steps and large propagation distances," IEEE Photon. Technol. Lett., **3**, 466-468 (1991).

[342] W.P. Huang, S.T. Chu and S.K. Chaudhuri, "A semivectorial finite-difference time-domain method," IEEE Photon. Technol. Lett., **3**, 803-806 (1991).

[343] W.P. Huang, C.L. Xu, S.T. Chu and S.K. Chaudhuri, "A vector beam propagation method for guided-wave optics," IEEE Photon. Technol. Lett., **3**, 910-913 (1991).

[344] H.E. Hernandez-Figueroa, "Nonlinear nonparaxial beam-propagation method," Electron. Lett., **30**, 352-353 (1994).

[345] U. Hempelmann and L. Bersiner, "Wave propagation in acoustooptical anisotropic waveguides," IEE Proc., Part J, **140**, 193-200 (1993).

[346] G.P. Argawal, "Induced focusing of optical beam in self-defocusing nonlinear media," Phys. Rev. Lett., **64**, 2487-2490 (1990).

[347] A. Chatterjee and J.L. Volakis, "Conformal absorbing boundary condition for the vector wave equation," Microwave Opt. Technol. Lett., **6**, 886-889 (1993).

[348] G.R. Hadley, "Transparent boundary condition for beam propagation," Opt. Lett., **16**, 624-626 (1991).

[349] W.D. Smith, "A nonreflecting plane boundary for wave propagation problems," J. Comput. Phys., **15**, 493-503 (1974).

[350] H.A. Haus, "Molding light into solitons," IEEE Spectrum, March 1993.

[351] T.E. Bell, "Light that acts like 'natural bits'," IEEE Spectrum, August 1990.

[352] G.P. Agrawal, **Nonlinear fiber optics**, San Diego: Academic Press, 1989.

[353] N.J. Doran, "Solitons: permanent waves," IEE Review September 1992.

[354] M. Eguchi, K. Hayata and M. Koshiba, "Analysis of soliton pulse propagation in an optical fiber using the finite-element method," Electron. Commun. Japan, Part 2, **73**, 81-91 (1990).

[355] M.W. Chbat *et al.*, "Ultrafast soliton-trapping and gate," J. Lightwave Technol., **10**, 2011-2016 (1992).

[356] R.H. Enns, R. Fung and S.S. Rangnekar, "Optical crosstalk-induced switching between bistable soliton states," IEEE J. Quantum Electron., **27**, 252-258 (1991).

[357] M.N. Islam, "Ultrafast all-optical logic gates based on soliton trapping in fibers," Opt. Lett., **14**, 1257-1259 (1989).

[358] Q. Wang, P.K. Wai, C.J. Chen and C.R. Menyuk, "Numerical modeling of soliton-dragging logic gates," J. Opt. Soc. Am. B, **10**, 2030-2039 (1993).

[359] K. Tamura, L.E. Nelson, H.A. Haus and E.P. Ippen, "Soliton versus nonsoliton operation of fiber ring lasers," Appl. Phys. Lett., **64**, 149-151 (1994).

[360] F. Fontana *et al.*, "Self-starting sliding frequency fibre soliton laser," Electron. Lett., **30**, 321-322 (1994).

[361] A.B. Grudinin, D.J. Richarson and D.N. Payne, "Passive harmonic modelocking of a fibre soliton ring laser," Electron. Lett., **29**, 1860-1861 (1993).

[362] W. Zhao and E. Bourkoff, "Propagation properties of dark solitons," Opt. Lett., **14**, 703-705 (1989).

[363] Y. Chen, "Vector dark spatial solitons," Electron. Lett., **27**, 1346-1348 (1991).

[364] Y. Kodama, M. Romagnoli and S. Wabnitz, "Stabilisation of optical solitons by an acousto-optic modulator and filter," Electron. Lett., **30**, 261-262 (1994).

[365] Y. Silberberg, "Collapse of optical pulses," Opt. Lett., **15**, 1282-1284 (1990).

[366] Y. Silberberg, J.S. Aitchison, A.M. Weiner *et al.*, "Self-focusing revisited: spatial solitons, light bullets and optical pulse collapse," **Directions in Electromagnetic Wave Modelling** [Proc. Int. Conf., eds: H.L. Bertoni and L.B. Felsen], New York: Plenum, 529 (1991).

[367] N.N. Akhmediev and R.F. Nabiev, "Modulation instability of the ground state of cylindrical waveguide: optical machine-gun," Proceedings of 16th Australian Conference on Optical Fibre Technology, IREE Australia, 310-313 (1991).

Index

LICENSE AGREEMENT

Hardware and Software Requirements

Because the software is supplied in standard FORTRAN 77 source code, it may be run on any computer with a good FORTRAN 77 compiler installed. When intensive computation is involved in computing higher-order modes and or nonlinear structures, a workstation or a mainframe computer is highly recommended. The actual RAM and hard disk spaces required depend on the parameters used. Typically, 4–8 megabytes of RAM and a hard disk with 5 megabytes of free space would be sufficient.

Printed and bound by CPI Group (UK) Ltd, Croydon, CR0 4YY
24/10/2024
01778291-0004